Shikha Kaushik, Anju Singh
Biomolecules

Also of Interest

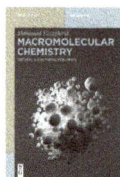

Macromolecular Chemistry.
Natural and Synthetic Polymers
Mohamed Elzagheid, 2021
ISBN 978-3-11-076275-4, e-ISBN 978-3-11-076276-1

Bioinorganic Chemistry.
Some New Facets
Ram Charitra Maurya, 2021
ISBN 978-3-11-072729-6, e-ISBN 978-3-11-072730-2

Chemistry of Nucleic Acids
Harri Lönnberg, 2020
ISBN 978-3-11-060927-1, e-ISBN 978-3-11-060929-5

Protein Chemistry
Lars Backman, 2019
ISBN 978-3-11-056616-1, e-ISBN 978-3-11-056618-5

Organic Chemistry.
Fundamentals and Concepts, 2nd edition
John M. McIntosh, 2022
ISBN 978-3-11-077820-5, e-ISBN 978-3-11-077831-1

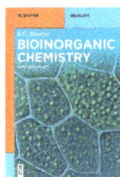

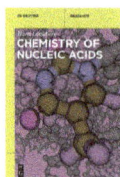

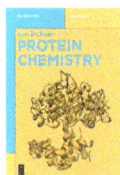

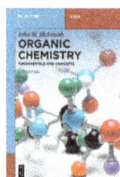

Shikha Kaushik, Anju Singh

Biomolecules

From Genes to Proteins

DE GRUYTER

Authors
Dr. Shikha Kaushik
Department of Chemistry
Rajdhani College
University of Delhi
110015 New Delhi
India
shikha.kaushik@rajdhani.du.ac.in

Dr. Anju Singh
Department of Chemistry
Ramjas College
University of Delhi
110007 New Delhi
India
anjusingh@ramjas.du.ac.in

ISBN 978-3-11-079375-8
e-ISBN (PDF) 978-3-11-079376-5
e-ISBN (EPUB) 978-3-11-079415-1

Library of Congress Control Number: 2022949442

Bibliographic information published by the Deutsche Nationalbibliothek
The Deutsche Nationalbibliothek lists this publication in the Deutsche Nationalbibliografie; detailed
bibliographic data are available on the Internet at http://dnb.dnb.de

© 2023 Walter de Gruyter GmbH, Berlin/Boston
Cover image: peterschreiber.media/iStock/Getty Images
Typesetting: Integra Software Services Pvt. Ltd.
Printing and binding: CPI books GmbH, Leck

www.degruyter.com

Preface

The internal machinery of life, the chemistry of the parts, is something beautiful. And it turns out that all life is interconnected with all other life. –Richard P. Feynman

Biochemistry is an important and interesting subdiscipline comprising both chemistry and biology dealing with the chemical processes occurring within living organisms. It is the study of the structures and functions of biological molecules such as nucleic acids, proteins, carbohydrates, and lipids. These biomolecules play an important role in various biochemical processes that occur in cells and organisms; by studying them one can understand their significance in various cellular processes that regulate the growth and development of living organisms.

A long span of interaction with students has guided us to give special attention to making difficult topics simple and comprehensible. *Introduction to Biomolecules* has been written to cater to the needs of undergraduate students of all Indian universities. The book is structured into five chapters, which cover the core concepts and fascinating chemistry of biomolecules (i.e., nucleic acids, proteins, enzymes, carbohydrates, and lipids), and concepts of energy in biosystems. The contents of the book will help students in acquiring in-depth knowledge and information pertinent to the fundamental aspects of biomolecules. This book also exposes students to different metabolic pathways and the transfer of energy between molecules that function within the biological systems. We truly believe that this book contains all that is needed to understand the topics systemically and shall prove to be a valuable piece of material for the students of chemistry, biochemistry, biotechnology, and biosciences, aiming to enhance the knowledge of the reader.

Our first and foremost gratitude goes to Prof. Shrikant Kukreti (Senior Professor, Department of Chemistry, University of Delhi) for his kind guidance and valuable suggestions during the preparation of the manuscript. His thoughtful encouragement, knowledge, and continual support have made this endeavor possible. We are extremely grateful to Prof. Rajesh Giri (Principal, Rajdhani College, University of Delhi) for his endless support and understanding spirit that kept us motivated. The authors would like to thank Prof. Manoj Kumar Khanna (Principal, Ramjas College, University of Delhi) for his kind cooperation and invaluable help. Thanks are also due to our dear students and worthy colleagues who helped in one way or other. Our warm and most earnest thanks go to our families for their kind patience and unfailing support and cooperation during the entire work span. The authors would love to mention special thanks to our dear Akshita, Advvay, and Master Yatharth for their peaceful smiles and tender love which kept us going. Last, but not least, we would like to acknowledge with gratitude the support of the publishing team at esteemed De Gruyter.

All possible care and efforts have been taken to make the book error free; however, no work is perfect. We would welcome and be pleased to receive suggestions

https://doi.org/10.1515/9783110793765-202

from the teaching fraternity and students and general readers for the improvement of subsequent editions.

Shikha Kaushik

Anju Singh

Contents

Chapter 2
Amino Acids, Peptides, and Proteins —— 59

About the Authors

Dr. Shikha Kaushik is Assistant Professor (chemistry) at Rajdhani College, University of Delhi. She received her bachelor's degree in chemistry, and master's degree in organic chemistry. She did her Ph.D. in biophysical chemistry from the University of Delhi. She has been teaching graduate students of B.Sc. (H) chemistry and other courses for more than 12 years. Dr. Kaushik was awarded Research Fellowship in Science for Meritorious Students (RFSMS) by UGC for pursuing her research work. She has published research articles in peer-reviewed journals of international repute.

Dr. Anju Singh is Assistant Professor (chemistry) at Ramjas College, University of Delhi. She is M.Sc. in organic chemistry and did her Ph.D. in biophysical chemistry from the University of Delhi. Dr. Anju has more than 8 years of teaching experience. She has been teaching undergraduate students of B.Sc. (H) chemistry and B.Sc. (P) life sciences. She is also actively involved in research and has published research papers in peer-reviewed journals of international repute.

https://doi.org/10.1515/9783110793765-204

Chapter 1
Carbohydrates and Lipids

Carbohydrates are naturally occurring biological molecules and are one of the three macronutrients of our diet along with proteins and fats. The main function of carbohydrates is to provide instant energy. They also play a crucial role in the structure and function of cells, tissues, and organs, as well as in the formation of carbohydrate structures on the surface of cells. Monosaccharide, specifically glucose, is a vital source of energy for virtually all forms of life. In addition to this, it is also utilized in other metabolic pathways such as for the biosynthesis of glycogen and lipids. Fructose, another monosaccharide, is the most used sweetener across the world. In the cells, fructose is converted into glucose derivatives, and glucose, which is not immediately used by the cells, is stored in the liver and muscle cells, and this storage form is known as glycogen. In plants, it is stored in the form of starch. Carbohydrates are abundantly found as cellulose, which is a structural component of the cell wall in various plants and algae. Ribose, a pentose sugar, is an important structural component of nucleic acids; deoxyribose is present in deoxyribonucleic acid (DNA), whereas ribose is present in ribonucleic acid (RNA). Ribulose and xylulose are found in pentose phosphate pathway, whereas galactose, a component of milk sugar lactose, occurs in galactolipids which is a constituent of plant cell membranes and glycoproteins in many tissues. Mannose plays an important role in human metabolism especially in glycosylation of certain proteins. Trehalose, a natural occurring sugar, consists of two molecules of glucose and acts as an energy source in many organisms. It also functions as a protectant against the effects of dehydration and freezing.

Carbohydrates are very important components of biological system and they are composed of carbon (C), hydrogen (H), and oxygen (O), in which H and O are present in the ratio 2:1. They are known as hydrates of carbon. Carbohydrates can also be referred to as polyhydroxy aldehydes or ketones. In biochemistry, it is used as a synonym of saccharide which includes sugars, starch, and cellulose. The word "saccharide" is originated from the *Greek* word *sákkharon*, which means sugar, and can be categorized into four main groups: monosaccharides, disaccharides, oligosaccharides, and polysaccharides.

1.1 Classification of Carbohydrates

Carbohydrates are classified into four main classes:

i. Monosaccharides: Monosaccharides, also known as simple sugars, are carbohydrates which contain polyhydroxy aldehyde (-CHO) and ketone (>C = O) groups. The general formula of monosaccharides is $C_n(H_2O)_n$. These sugars cannot be further hydrolyzed into simpler compounds.

https://doi.org/10.1515/9783110793765-001

Monosaccharides can be further subclassified on the basis of
(a) the number of carbon atoms in the chain – a monosaccharide containing three carbon atoms is named as triose ($C_3H_6O_3$), four carbon atoms is tetrose ($C_4H_8O_4$), five carbon atoms is pentose ($C_5H_{10}O_5$), and six carbon atoms is hexose ($C_6H_{12}O_6$).
(b) carbonyl functional group in the saccharide – monosaccharides which contain an aldehyde group (-CHO) are called aldoses, whereas those containing a ketone group ($>C = O$) are termed as ketoses.

Both the parameters, that is, the number of carbon atoms and the type of carbonyl group, are considered to classify a monosaccharide. For example, the most important pentoses include deoxyribose (a component of DNA), ribose (present in RNA and several vitamins), and xylose (found in woody materials in the form of xylan). Glucose, mannose, and galactose are among the most common dietary aldohexoses, whereas fructose is the most abundant ketohexose found in nature.

Glucose, considering all its combined forms, is the most abundant monosaccharide present in nature, and is found in fruits, vegetables, dried fruits, and honey. Galactose is usually found in nature in combined form with other sugars. For example, it is present in milk in combination with glucose as lactose (milk sugar). Fructose is the only important ketohexose that is present in free form in fruits and honey along with glucose.

ii. Disaccharides: Disaccharides, also known as double sugars, are formed when two monosaccharides (simple sugars) are joined covalently via O-glycosidic linkage. In other words, disaccharides are the condensation products of two same or different monosaccharides. The general formula of disaccharides is $C_n(H_2O)_{n-1}$. Sucrose is the most common disaccharide composed of one glucose and one fructose molecule. It occurs predominantly in sugar cane and sugar beets, with smaller amounts found in fruits, vegetables, nuts, and honey. Other common disaccharides are lactose and maltose. Lactose is found naturally in milk and dairy products and consists of galactose and glucose, whereas maltose is made up of two molecules of glucose and is found in malt from germinating grains such as barley.

iii. Oligosaccharides: Oligosaccharides refer to carbohydrates that contain between three and ten monosaccharide residues, joined by a glycosidic bond. However, some authors also classified carbohydrates with up to 20 residues. The relationship of monosaccharides to oligosaccharides or polysaccharides is similar to that of nucleotides to nucleic acids and amino acids to proteins. The most common oligosaccharides are raffinose, stachyose, and verbascose. Raffinose ($C_{18}H_{32}O_{16}$) is a trisaccharide consists of glucose, fructose, and galactose; stachyose ($C_{24}H_{42}O_{21}$), a tetrasaccharide, is composed of glucose, fructose, and two galactoses, whereas verbascose ($C_{30}H_{52}O_{26}$) is a pentasaccharide that is made up of glucose, fructose, and three galactoses. Many oligosaccharides can be found naturally in many fruits, vegetables, grains, and legumes. However, the majority of these cannot be broken down by the human digestive tract as we lack the digestive enzyme, *α-galactosidase*. Instead, they pass undigested into the lower gut where they are fermented and metabolized by anaerobic (beneficial) bacteria.

iv. Polysaccharides: These are ubiquitous long-chain polymeric carbohydrates that occur widely in nature, and consist of monosaccharide units linked together by glycosidic linkages. Polysaccharides can be composed of same (homopolysaccharide) or different (heteropolysaccharide) monosaccharide units, or can be linear or highly branched. Some of the most common polysaccharides are starch, cellulose, glycogen, chitin, and pectins.

1.2 Relative Configuration of Sugars: D- and L-Sugars

1.2.1 Monosaccharides

As mentioned earlier, monosaccharides are also known as polyhydroxy aldoses (polyhydroxy) or polyhydroxy ketoses (polyhydroxy ketones). Based on the number of carbon atoms present in the structure, these can be further subclassified into trioses, tetroses, pentoses, and so on. All naturally occurring carbohydrates have chiral carbon and are therefore known as optically active polyhydric aldehydes or ketones.

Glyceraldehyde is the parent compound of the class of monosaccharides known as trioses (aldoses), and has an asymmetric carbon (marked by *), which exists in two enantiomeric forms. According to the Fischer projection, when -OH group is present on the right side in an enantiomer, it is referred to as D-glyceraldehyde; and when -OH group is on the left side, it is called L-glyceraldehyde. D-Glyceraldehyde was chosen as an arbitrary standard by Emil Fisher to assign the configuration of higher carbohydrates. Besides carbohydrates, the configurations of many other important biological compounds have been related to glyceraldehyde such as α-amino acids, steroids, and terpenes:

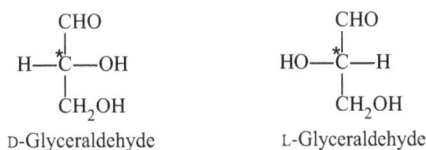

D-Glyceraldehyde   L-Glyceraldehyde

The maximum number of possible optical isomers of carbohydrates can be calculated by knowing the number of asymmetric carbons in the molecule.

Maximum number of optical isomers is 2^n, where n is the number of chiral (asymmetric) carbons.

[*Note*: Dihydroxyacetone ($HOCH_2.CO.CH_2OH$) is known as ketotriose but it does not belong to carbohydrate because it is not optically active.]

One of the enantiomers undergoes a series of complex reactions resulting in the synthesis of all aldoses. The overall method involves adding carbons one at a time, and hence, aldotetroses can be prepared from glyceraldehyde; the aldotetroses can then be used to prepare aldopentoses. Further, aldopentoses can be used to synthesize aldohexoses, and similarly other higher aldoses can be synthesized. When glyceraldehyde stepped up to give an aldotetrose, the carbon of the aldehyde group of glyceraldehyde is reduced, thus, generating a new asymmetric carbon. The carbon atom added during the reaction then becomes the aldehyde group of the newly synthesized aldotetrose:

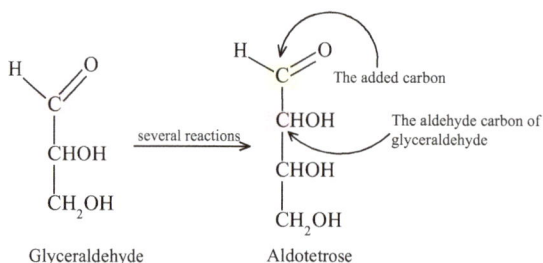

Two aldotetroses are prepared from D-glyceraldehyde and two from L-glyceraldehyde, resulting in the generation of four aldotetroses. The structures and relationships of D- and L-glyceraldehyde to the four possible aldotetroses are shown in Figure 1.1.

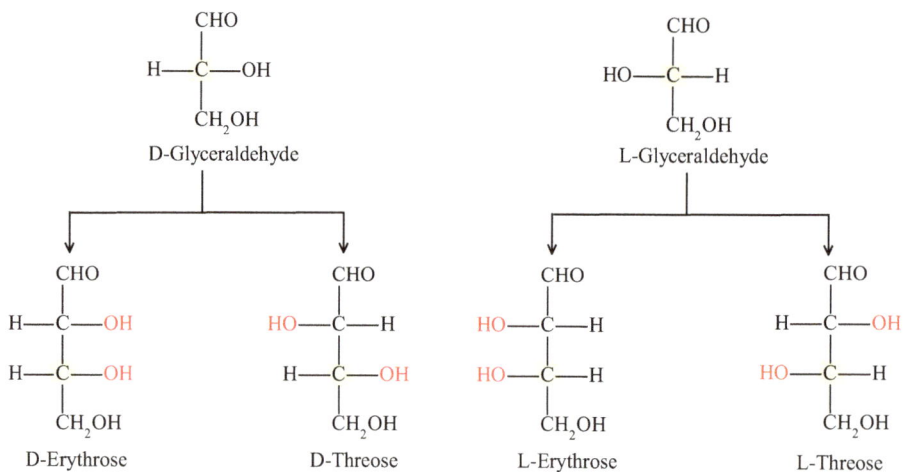

Figure 1.1: Four aldotetroses and their relationship with enantiomeric glyceraldehydes.

It is pertinent to mention that the chiral (asymmetric) carbon of glyceraldehyde now becomes the asymmetric carbon that is farthest from the -CHO group in newly synthesized aldoses (shown in yellow in Figure 1.1). We can extend the aldose family starting from glyceraldehyde, and all aldoses in which the -OH group on the chiral (asymmetric) carbon farthest from the aldehyde group (-CHO) on the right side are related to D-glyceraldehyde and designated as D-sugars; and the aldoses which have -OH group on the asymmetric carbon farthest from the -CHO group on the left are referred to as L-sugars. This is termed as relative configuration of sugars.

In a similar manner, ketotetrose has one asymmetric carbon and exists as enantiomeric pairs, known as D- and L-erythrulose:

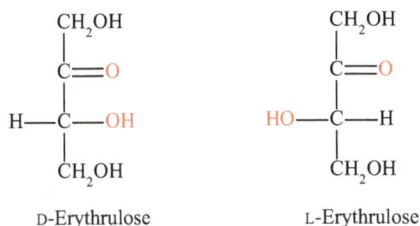

CH₂OH structures:

D-Erythrulose — CH$_2$OH / C=O / H—C—OH / CH$_2$OH

L-Erythrulose — CH$_2$OH / C=O / HO—C—H / CH$_2$OH

D-Erythrulose   L-Erythrulose

Pentoses, another important group of monosaccharides, contain three asymmetric carbon atoms (2, 3, 4 carbons), and therefore, exist in eight optically active forms (2^3). It is interesting to know that that only D-arabinose, D-ribose, D-xylose, and L-arabinose are found in nature, and rest are synthetic pentoses. 2-Deoxyribose is a pentose sugar found in the nucleic acids. The prefix "deoxy" implies the replacement of a hydroxy group by hydrogen at C-2.

Aldohexoses are the most important group of monosaccharides. They contain four structurally different asymmetric (chiral) carbon atoms, and therefore can exist in 16 optically active forms. All forms belong to D- and L-forms of glucose, galactose, mannose, altrose, allose, gulose, iodose, and talose. It has been found that almost all the monosaccharides found in nature generally belong to the D-sugar family, so we will consider here only those belonging to the D-series. The D-family of aldoses is shown in Figure 1.2.

1.3 D-(+)-Glucose/Dextrose (Grape Sugar): Structure and Properties

D-Glucose, considering all its combined forms, is the most common and abundant naturally occurring monosaccharide. Glucose was isolated from raisins by Andreas Margraf in 1747, while the name "glucose" was coined by Jean Dumas in 1838 and is derived from the Greek word *glycos* meaning sugar or sweet. It is a white crystalline solid which is highly soluble in water. It is found in ripe grapes to an extent of 20%, and hence known as grape sugar. It is also found in honey and sweet fruits. Glucose is also named as dextrose because it exists in nature mainly as optically active dextrorotatory isomer. Glucose is an essential constituent of blood, and the expected values for normal fasting blood glucose are between 70 and 100 mg/dL but in diabetic persons the level may be much higher.

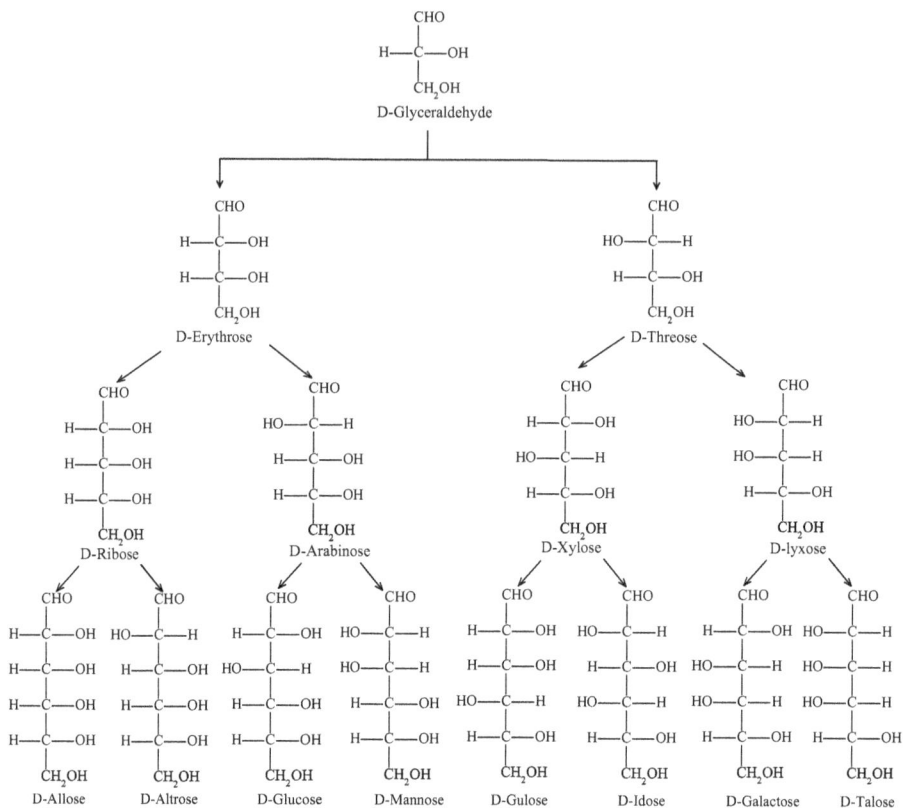

Figure 1.2: D-family of aldoses.

1.3.1 Preparation of Glucose

1. From starch

Commercially, glucose is prepared by hydrolysis of starch by boiling it in acidic medium at high temperature and pressure:

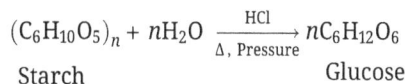

$$(C_6H_{10}O_5)_n + nH_2O \xrightarrow[\Delta, \text{Pressure}]{\text{HCl}} nC_6H_{12}O_6$$

Starch   Glucose

2. From sucrose (cane sugar)

Glucose can be prepared in the laboratory by boiling sucrose (cane sugar) with dilute hydrochloric acid or sulfuric acid for about 2 h. This reaction yields glucose and fructose in equal amount. Glucose can be separated from fructose by the addition of alcohol during cooling. Glucose is almost insoluble in alcohol, and crystallizes out first, while fructose being more soluble, it remains in the solution. The solution is then filtered to get the crystals of glucose:

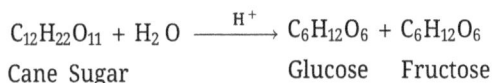

$$C_{12}H_{22}O_{11} + H_2O \xrightarrow{H^+} C_6H_{12}O_6 + C_6H_{12}O_6$$

Cane Sugar   Glucose   Fructose

1.3.2 Structure Elucidation of D(+)-Glucose

The open-chain structure of glucose is derived on the basis of following facts:

(i) Elemental analysis and molecular weight determination studies have revealed that the molecular formula of glucose is $C_6H_{12}O_6$.

(ii) Glucose on reduction with H_2/Ni produces sorbitol ($C_6H_{14}O_6$), which on treatment with hydrogen iodide converts to *n*-hexane. This reaction indicates that the six carbon atoms in a glucose molecule are present in a straight chain:

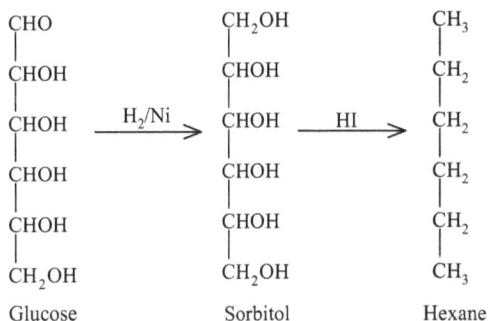

```
CHO                    CH₂OH                  CH₃
|                      |                      |
CHOH                   CHOH                   CH₂
|           H₂/Ni      |            HI        |
CHOH      ──────→      CHOH      ──────→      CH₂
|                      |                      |
CHOH                   CHOH                   CH₂
|                      |                      |
CHOH                   CHOH                   CH₂
|                      |                      |
CH₂OH                  CH₂OH                  CH₃
Glucose                Sorbitol               Hexane
```

(iii) Glucose on reaction with hydroxylamine leads to the formation of an oxime; also, it forms glucose phenylhydrazone when reacts with 1 mole of phenylhydrazine. These reactions indicate the presence of a carbonyl group in glucose:

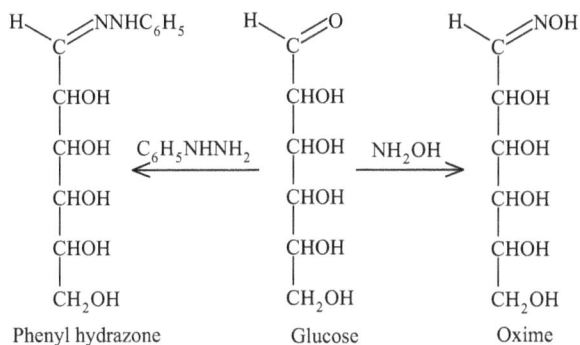

```
H    NNHC₆H₅        H      O          H      NOH
 \  /               \     /            \    /
  C                  C                   C
  |                  |                   |
CHOH               CHOH                CHOH
|                  |                   |
CHOH  C₆H₅NHNH₂    CHOH   NH₂OH        CHOH
|     ←──────      |      ──────→      |
CHOH               CHOH                CHOH
|                  |                   |
CHOH               CHOH                CHOH
|                  |                   |
CH₂OH              CH₂OH               CH₂OH
Phenyl hydrazone   Glucose            Oxime
```

(iv) Glucose reacts with 1 mol of hydrogen cyanide (HCN) to yield cyanohydrin. Cyanohydrin, on hydrolysis, followed by reaction with hydrogen iodide, yields heptanoic acid. This clearly indicates that the carbonyl group present in glucose is an aldehyde:

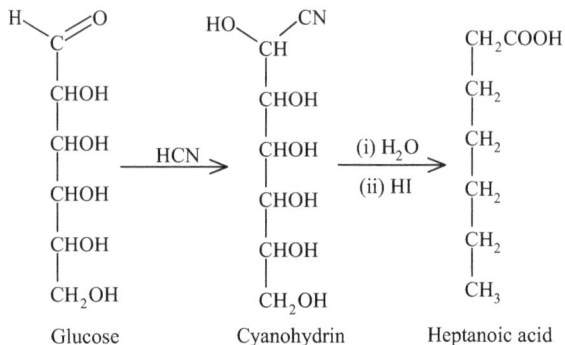

$$
\begin{array}{ccccc}
\text{H--C==O} & & \text{HO--CH--CN} & & \text{CH}_2\text{COOH} \\
\text{CHOH} & & \text{CHOH} & & \text{CH}_2 \\
\text{CHOH} & \xrightarrow{\text{HCN}} & \text{CHOH} & \xrightarrow[\text{(ii) HI}]{\text{(i) H}_2\text{O}} & \text{CH}_2 \\
\text{CHOH} & & \text{CHOH} & & \text{CH}_2 \\
\text{CHOH} & & \text{CHOH} & & \text{CH}_2 \\
\text{CH}_2\text{OH} & & \text{CH}_2\text{OH} & & \text{CH}_3 \\
\text{Glucose} & & \text{Cyanohydrin} & & \text{Heptanoic acid}
\end{array}
$$

(v) Glucose, on treatment with mild oxidizing agent such as bromine water, gives a monocarboxylic acid $C_6H_{12}O_7$, known as gluconic acid.

(vi) Glucose when treated with strong oxidizing agents such as conc. nitric acid gives glucaric acid ($C_6H_{10}O_8$, a dicarboxylic acid). Nitric acid oxidizes both the aldehyde group and the primary alcoholic group of an aldose to carboxylic acid:

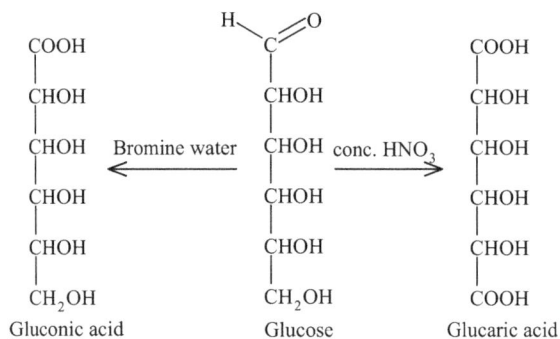

$$
\begin{array}{ccccc}
\text{COOH} & & \text{H--C==O} & & \text{COOH} \\
\text{CHOH} & & \text{CHOH} & & \text{CHOH} \\
\text{CHOH} & \xleftarrow{\text{Bromine water}} & \text{CHOH} & \xrightarrow{\text{conc. HNO}_3} & \text{CHOH} \\
\text{CHOH} & & \text{CHOH} & & \text{CHOH} \\
\text{CHOH} & & \text{CHOH} & & \text{CHOH} \\
\text{CH}_2\text{OH} & & \text{CH}_2\text{OH} & & \text{COOH} \\
\text{Gluconic acid} & & \text{Glucose} & & \text{Glucaric acid}
\end{array}
$$

(vii) Periodic acid (HIO_4) cleaves the carbon–carbon bonds containing -OH groups through oxidation. Glucose reacts with 5 mol of HIO_4 to give formaldehyde (1 mol) and formic acid (5 mol):

$$
\begin{array}{c}
\text{H} \diagdown \text{C} \diagup \!\!\!\!= O \\
| \\
\text{CHOH} \\
| \\
\text{CHOH} \\
| \\
\text{CHOH} \\
| \\
\text{CHOH} \\
| \\
\text{CH}_2\text{OH}
\end{array}
\quad \xrightarrow{\text{5 HIO}_4} \quad \text{5 HCOOH} + \text{HCHO}
$$

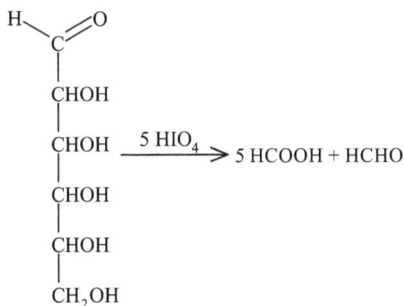

(viii) Glucose reduces Tollens' reagent (ammoniacal silver nitrate solution) and Fehling's solution, which confirms the presence of aldehyde group in glucose:

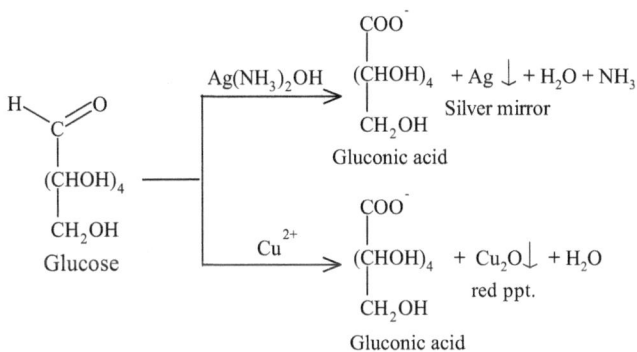

$$
\begin{array}{c}
\text{H} \diagdown \text{C} \diagup \!\!\!\!= O \\
| \\
(\text{CHOH})_4 \\
| \\
\text{CH}_2\text{OH} \\
\text{Glucose}
\end{array}
$$

$\xrightarrow{\text{Ag(NH}_3)_2\text{OH}}$
$$
\begin{array}{c}
\text{COO}^- \\
| \\
(\text{CHOH})_4 \\
| \\
\text{CH}_2\text{OH}
\end{array}
+ \text{Ag} \!\downarrow + \text{H}_2\text{O} + \text{NH}_3
$$
Silver mirror

Gluconic acid

$\xrightarrow{\text{Cu}^{2+}}$
$$
\begin{array}{c}
\text{COO}^- \\
| \\
(\text{CHOH})_4 \\
| \\
\text{CH}_2\text{OH}
\end{array}
+ \text{Cu}_2\text{O} \!\downarrow + \text{H}_2\text{O}
$$
red ppt.

Gluconic acid

(ix) Glucose, when reacts with acetic anhydride, produces a penta-acetyl derivative which confirms the presence of five hydroxy groups. The higher stability of the derivative indicates the presence of five hydroxy groups on different carbon atoms:

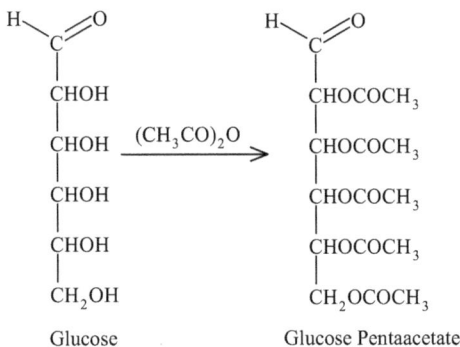

$$
\begin{array}{c}
\text{H} \diagdown \text{C} \diagup \!\!\!\!= O \\
| \\
\text{CHOH} \\
| \\
\text{CHOH} \\
| \\
\text{CHOH} \\
| \\
\text{CHOH} \\
| \\
\text{CH}_2\text{OH} \\
\text{Glucose}
\end{array}
\quad \xrightarrow{(\text{CH}_3\text{CO})_2\text{O}} \quad
\begin{array}{c}
\text{H} \diagdown \text{C} \diagup \!\!\!\!= O \\
| \\
\text{CHOCOCH}_3 \\
| \\
\text{CHOCOCH}_3 \\
| \\
\text{CHOCOCH}_3 \\
| \\
\text{CHOCOCH}_3 \\
| \\
\text{CH}_2\text{OCOCH}_3 \\
\text{Glucose Pentaacetate}
\end{array}
$$

Based on these reactions, it was concluded that glucose is a six-carbon, straight-chain pentahydroxy aldehyde (i.e., contains -CHO, -CHOH, and -CH$_2$OH groups), and the open-chain structure of glucose can be written as follows:

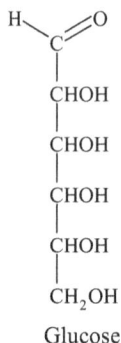

$$\underset{\text{Glucose}}{\overset{\displaystyle H\diagdown \!\!\!\!_C\!\!\diagup^{\displaystyle O}}{\underset{\displaystyle CH_2OH}{\overset{\displaystyle |}{\underset{|}{\overset{|}{\underset{|}{\overset{|}{\underset{|}{\overset{\displaystyle CHOH}{|}}}}}}}}}$$

CHOH
CHOH
CHOH
CHOH
CH$_2$OH
Glucose

The various reactions of glucose discussed earlier are summarized in Figure 1.3.

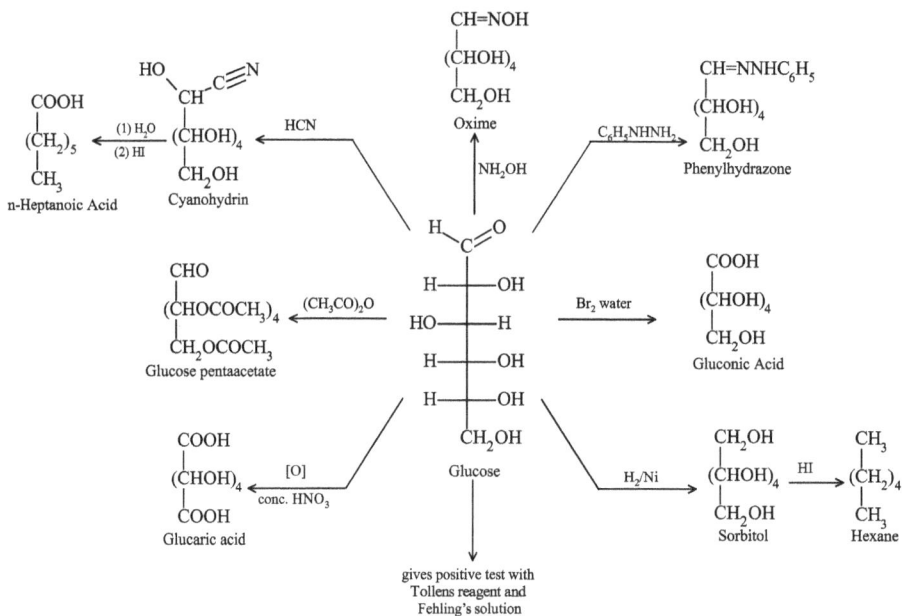

Figure 1.3: Reactions of glucose.

Carbonyl compounds, aldehydes, and ketones form hemiacetals and hemiketals, respectively, when react with alcohols. In monosaccharides, both the alcohol (-OH) group and the carbonyl (-C = O) group are present in the same molecules, thus, they undergo intramolecular reaction and form cyclic hemiacetal or hemiketals. Glucose, an aldohexose, forms an intramolecular hemiacetal which results from the reaction between the aldehyde group (carbon atom 1) and the hydroxy group (carbon atom 5). As a result of cyclization, a new chiral center is generated at C-1 (asymmetric), and the newly formed -OH group may be present either on the left or on the right in Fisher projection formulae. So, with the formation of hemiacetal, a mixture of two stereo-isomers which differ in the configuration of H and -OH groups around C-1 atom is produced. These two stereoisomers of D-(+)-glucose, known as α-D-glucose and β-D-glucose, are shown in Figure 1.4. The isomer having the -OH group on the right is called α-D-glucose and one having the -OH group on the left is called β-D-glucose. Such pairs of optical isomers which differ in the configuration only around C-1 atom are called anomers, and the asymmetric carbon at C-1 is known as anomeric carbon.

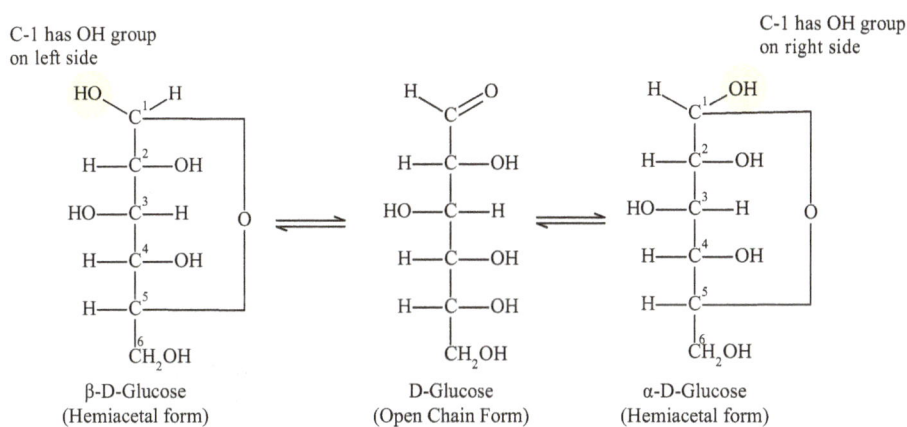

Figure 1.4: Structures of open-chain and cyclic hemiacetal forms of D-glucose (Fisher projection).

1.3.3 Mutarotation

Glucose is an optically active monosaccharide; when dissolved in water, it undergoes a change in its initial specific rotation till it attains a constant specific rotation. This gradual change in specific rotation is called mutarotation. D-(+)-Glucose exists in two enantiomeric forms, known as α-D-glucose and β-D-glucose. α-D-Glucose is obtained by crystallization of D-glucose from methanol, and its melting point is 146 °C. It has a specific rotation of +112.2°. The β-form of glucose, β-D-glucose, is obtained on crystallization of glucose from a hot saturated aqueous solution, and it melts at 150 °C. Its specific rotation is +18.7°.

When a solution of α-D-glucose or β-D-glucose in water is allowed to stand for some time, a gradual change in their specific rotation takes place till both achieves a value of +52.7°. The specific rotation of α-D-glucose decreases from +112.2° to +52.7° while that of β-D-glucose increases from +18.7° to +52.7°:

	α–D-Glucose	Equilibrium Mixture	β–D-Glucose
Specific Rotation	+112.2°	+52.7° (36 % α + 64 % β)	+18.7°
Melting point	146 °C		150 °C

Mutarotation

This alteration in specific rotation is termed as *mutarotation,* and represents an equilibrium mixture that comprises both α- and β-forms. Mutarotation involves the interconversion of the cyclic hemiacetals to open-chain form (aldehyde form) in solution, and ring opening followed by recyclization can form either the α- or β-form. A dynamic equilibrium occurs among three forms, and the equilibrium mixture contains α-D-glucose (36%) and β-D-glucose (64%) with less than 0.01% of the open-chain form. In cells, the mutarotation of glucose is catalyzed by an enzyme called *mutarotase.*

1.3.4 Osazone Formation: Reaction of Glucose with Excess of Phenylhydrazine

Osazone formation is one of the widely used reactions in carbohydrate chemistry. Osazones are quite useful for establishing the stereochemical relationships in isomeric sugars. It has already been discussed that aldohexoses on reaction with one mole of phenylhydrazine results in the formation of corresponding phenylhydrazones. Upon treating the hydrazones with two additional equivalents of phenylhydrazine, all the three hexoses (D-glucose, D-mannose, and D-fructose) are converted to the same osazone (Figure 1.5).

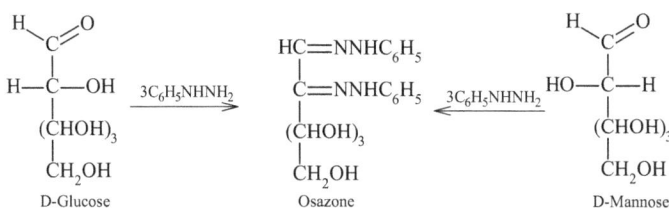

Figure 1.5: Osazone formation.

Emil Fischer first prepared osazone, and proposed the mechanism for formation of osazone; however, it had some limitations. Later, Weygand proposed an alternative mechanism of osazone formation that involves the Amadori rearrangement, which is discussed further:

Glucose

$\xrightarrow{C_6H_5NHNH_2}$

(with) Amadori rearrangement

$\xrightarrow{-C_6H_5NH_2}$

$\xrightarrow{C_6H_5NHNH_2}$

$\xrightarrow[-NH_3]{C_6H_5NHNH_2}$

Glucosazone

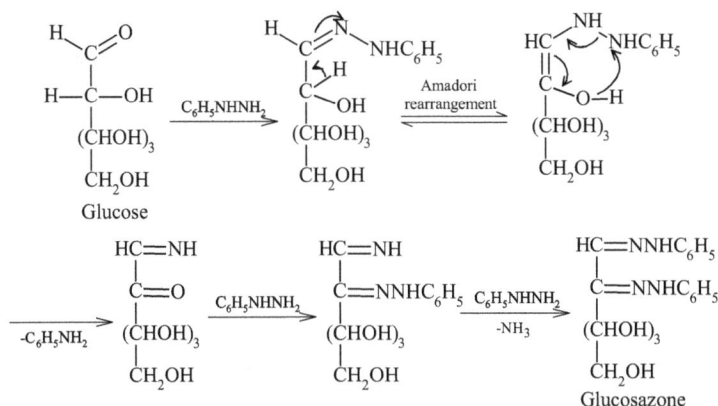

The first step in osazone formation is the reaction of aldehyde group of glucose with phenylhydrazine to form phenylhydrazone. The Amadori rearrangement takes place in the second step in which -CHOH group at C-2 is oxidized to keto group, followed by the elimination of aniline. The third step involves the reaction of newly generated carbonyl (keto) group with 2 mol of phenylhydrazine resulting in the formation of glucosazone.

Glucose and mannose are epimers, and they differ in their configuration only at C-2; they give same osazone on treatment with 3 mol of phenylhydrazine.

Now the question arises: why does the secondary alcoholic group at C-3 not react with phenylhydrazine? The reason is that the formed glucosazone is stabilized through chelation, and hence reaction stops beyond C-2:

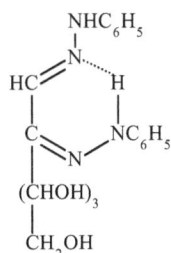

1.3.5 Cyclic Structure of D-(+)-Glucose: Formation of Glucosides

The open-chain structure of glucose accounted for most of the properties of glucose; however, it could not explain the following facts:
(i) D-Glucose fails to undergo some of the reactions typical of aldehydes; for example, it has an aldehyde group, but gives a negative test with Schiff's reagent.
(ii) It does not react with ammonia and sodium bisulfite to form addition products, which are characteristic tests for aldehydes.

(iii) Glucose reacts with hydroxylamine to form an oxime but glucose pentaacetate does react with hydroxylamine which indicates the absence of free aldehyde group in glucose.

(iv) D-(+)-glucose occurs in two isomeric forms, known as α-D-glucose and β-D-glucose, which undergo mutarotation, and it is possible only when glucose forms a cyclic structure. The open-chain structure does not support change in specific rotation and mutarotation.

(v) When glucose reacts with methanol in the presence of dilute hydrochloric acid, it forms two isomeric methyl glucosides (Figure 1.7), that is, methyl α-D-glucoside (melting point = 165 °C and specific rotation = +158°) and methyl β-D-glucoside (melting point = 107 °C and specific rotation = −33°). These glucosides do not give positive test with Tollens' reagent and Fehling's solution, and do not undergo mutarotation.

To fix all the above facts, a picture about the structure of D-(+)-glucose had to be changed. As a result of the pioneer work by many chemists including Fischer, Tollens, and Tanret, the cyclic structure of D-(+)-glucose emerged out in 1895. The ring size was elucidated, and the preferred conformation was confirmed. It was revealed that glucose exists not as an open-chain aldehyde structure but forms a cyclic "internal" hemiacetal structure which is best characterized by Haworth's projection as a pyranose ring. The hemiacetal structure formation results from the intramolecular nucleophilic addition of the -CHOH group at carbon atom 5 to the aldehyde group at carbon atom 1.

1.3.5.1 The Haworth Projection Formulae

The cyclic structure of glucose was proposed by Haworth, and the Haworth formulae is a better way of representing the hemiacetal forms of sugars as Fischer projection formulae do not portray the cyclic hemiacetal forms of sugars very well.

In Haworth structures, drawn with the heterocyclic oxygen in the upper right corner, all groups written to the left of the Fischer projection are written upward, whereas all groups written on the right of the Fischer projection are written downward. The α-form has the –OH group on C-1 pointing "down," whereas the β-form has the –OH group on C-1 pointing "up." The cyclic structures of α-D-glucose and β-D-glucose, and methyl α-D-glucoside and methyl β-D-glucoside are shown in Figures 1.6 and 1.7, respectively.

Glucose anomers: Hemiacetals Reducing sugars

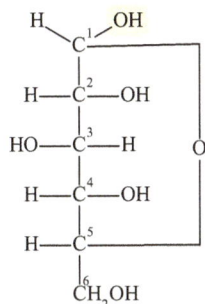 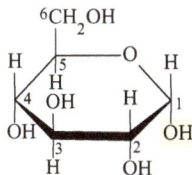 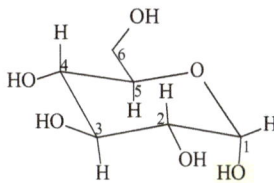

α-D-(+)-Glucose (m.p. 146 °C, [α] = + 112.2°)

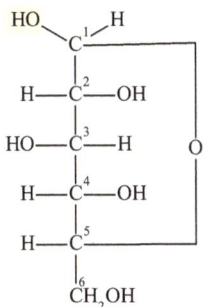 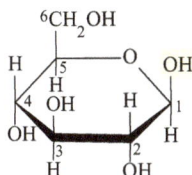 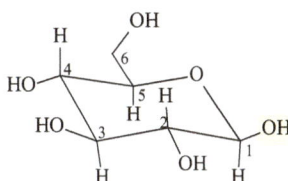

β-D-(+)-Glucose (m.p. 150 °C, [α] = + 18.7°)

Figure 1.6: Cyclic structures of D-(+)-glucose.

Glucose anomers: Acetals Non-Reducing sugars

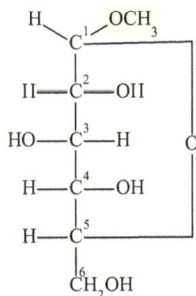 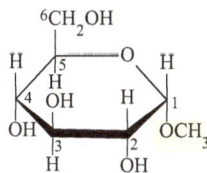 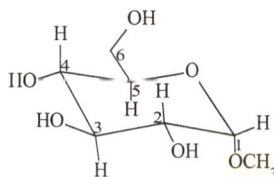

Methy α-D-(+)-Glucoside (m.p. 165 °C, [α] = + 158°)

α

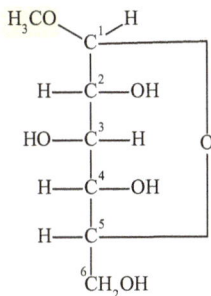 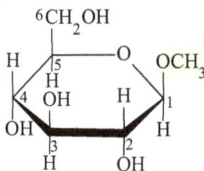 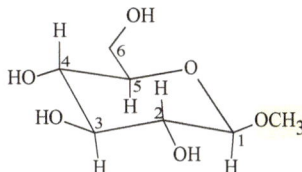

Methy β-D-(+)-Glucoside (m.p. 107 °C, [α] = - 33°)

Figure 1.7: Cyclic structures of methyl D-(+)-glucoside.

1.4 Lengthening the Carbon Chain of Aldoses: Kiliani–Fischer Synthesis

Aldose chains may be lengthened by one carbon atom using Kiliani–Fischer synthesis, and involves the following four steps:
(i)   Addition of HCN (formation of diastereomeric cyanohydrins)
(ii) Hydrolysis of -CN to -COOH (formation of two diastereomeric aldonic acid)
(iii) Conversion of aldonic acid to lactone by heating
(iv) Reduction of lactone with sodium amalgam (Na–Hg) to get higher aldose

Aldopentose when treated with HCN results in the formation of cyanohydrin. This step lengthens the parent carbon chain by one carbon atom, and generates a new asymmetric carbon, which results in the formation of two diastereomeric cyanohydrins. Further, cyanohydrins undergo hydrolysis along with subsequent acidification consequently producing diastereomeric aldonic acids. The acids undergo dehydration reaction and then reduction with sodium amalgam that results in the formation of two aldoses having one carbon atom more than the parent carbon. The two aldoses obtained by this method differ in their configuration at carbon atom 2. Such pairs of diastereomers that differ in stereochemistry at only one of many asymmetric centers are known as *epimers*. Figure 1.8 illustrates the conversion of a D-arabinose (aldopentose) to D-glucose and D-mannose (aldohexoses).

1.5 Shortening of Carbon Chain of Aldoses

Aldose chains may be shortened by one carbon atom by two methods: Ruff degradation and Wohl's degradation.

i. Ruff degradation: This is the most widely used method in which oxidation of higher aldose takes place which yields aldonic acid. Further treatment of aldonic acid with calcium carbonate gives calcium salt of aldonic acid. The calcium salt is then oxidized by Fenton's reagent (hydrogen peroxide in the presence of ferrous salts) which produces aldose containing one carbon atom less than the parent molecule. The conversion of D-glucose (aldohexose) to D-arabinose is shown in Figure 1.9.

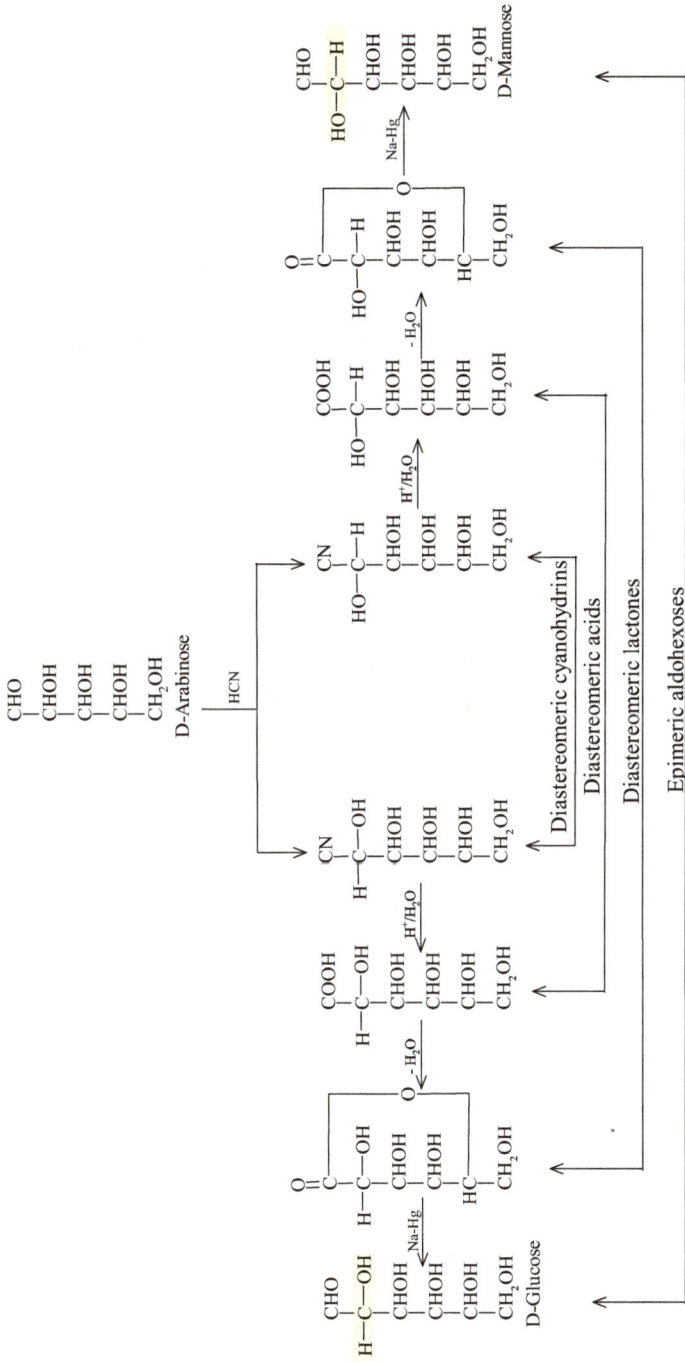

Figure 1.8: Kiliani–Fischer synthesis.

```
CHO                      COOH                    COO⁻]Ca²⁺                    CHO
|                        |                       |                            |
CHOH                     CHOH                    CHOH                         CHOH
|                        |                       |                            |
CHOH    Bromine water    CHOH     CaCO₃          CHOH     H₂O₂/Fe²⁺           CHOH
|       ──────────→      |       ──────→         |       ──────→             |
CHOH                     CHOH                    CHOH                         CHOH
|                        |                       |                            |
CHOH                     CHOH                    CHOH                         CH₂OH
|                        |                       |
CH₂OH                    CH₂OH                   CH₂OH

D-Glucose                Gluconic acid           Calcium salt of             D-Arabinose
                                                 gluconic acid
```

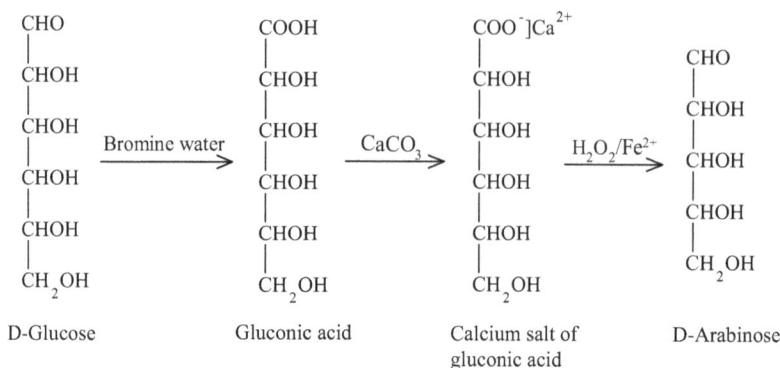

Figure 1.9: Ruff degradation (conversion of D-glucose to D-arabinose).

ii. Wohl's degradation: This is another method which is used to convert an aldose to another aldose with one carbon less than the parent aldose. Here, we have illustrated the conversion of D-glucose (aldohexose) to D-arabinose (aldopentose) using Wohl's degradation. First, D-glucose is converted into its oxime by treatment with hydroxylamine. The oxime when treated with acetic anhydride followed by dehydration yields polyacetylated nitrile. Further heating of the product with ammoniacal silver oxide yields D-arabinose (aldopentose). The degradation procedure takes place as shown in Figure 1.10.

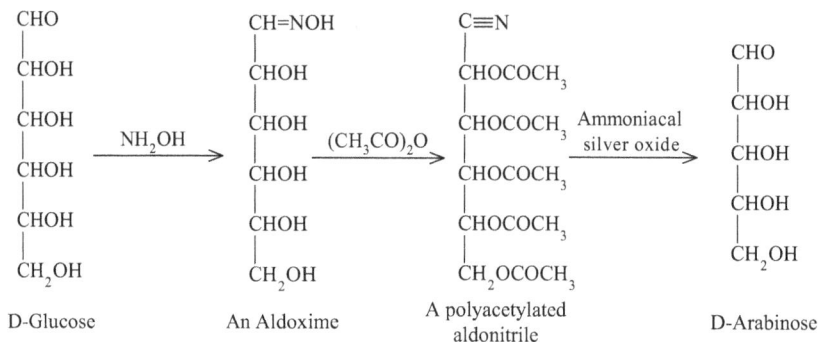

```
CHO                      CH=NOH                  C≡N                          CHO
|                        |                       |                            |
CHOH                     CHOH                    CHOCOCH₃                     CHOH
|                        |                       |            Ammoniacal      |
CHOH     NH₂OH           CHOH    (CH₃CO)₂O       CHOCOCH₃    silver oxide     CHOH
|       ──────→          |       ──────→         |          ──────→          |
CHOH                     CHOH                    CHOCOCH₃                     CHOH
|                        |                       |                            |
CHOH                     CHOH                    CHOCOCH₃                     CH₂OH
|                        |                       |
CH₂OH                    CH₂OH                   CH₂OCOCH₃

D-Glucose                An Aldoxime             A polyacetylated            D-Arabinose
                                                 aldonitrile
```

Figure 1.10: Wohl's degradation (conversion of D-glucose to D-arabinose).

1.6 Interconversion of D-Glucose, D-Mannose, and D-Fructose

The α-hydrogen present in aldoses and ketoses is acidic in nature, and is abstracted in the presence of an alkali, resulting in the generation of an enolate ion. The resultant enolate ion is stabilized by resonance. It undergoes protonation at different positions, yielding different products:

- if protonation occurs at carbon, it results in the formation of epimers known as epimeric aldoses, and
- if it takes place at oxygen, it yields ketose through ene–diol rearrangement

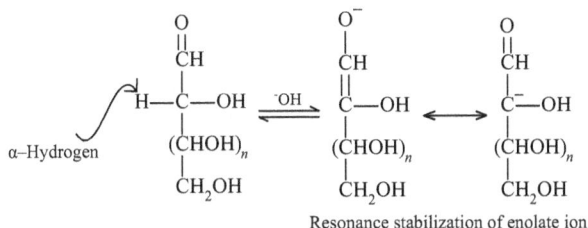

$$
\begin{array}{ccc}
\overset{O}{\underset{\parallel}{C}}H & \overset{O^-}{\underset{|}{C}}H & \overset{O}{\underset{\parallel}{C}}H \\
H-\overset{|}{C}-OH \;\;\xrightarrow{\;^-OH\;} & \overset{\parallel}{C}-OH \longleftrightarrow & \overset{-}{C}-OH \\
(CHOH)_n & (CHOH)_n & (CHOH)_n \\
CH_2OH & CH_2OH & CH_2OH
\end{array}
$$

α–Hydrogen

Resonance stabilization of enolate ion

(a) D-Glucose on treatment with an alkali loses α-hydrogen present at C-2 position, resulting in the generation of an enolate ion. Further, protonation regenerates the stereocenter at C-2 position that was lost during enolization process, and now the proton can be attached from either side to yield two isomeric forms of aldoses, that is, D-glucose and D-mannose. It is important to mention here that D-glucose and D-mannose are C-2 epimers, which means they differ in configuration at C-2 only:

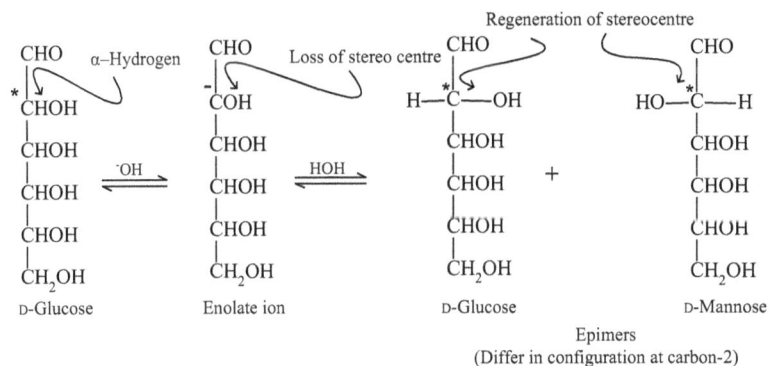

Regeneration of stereocentre

$$
\begin{array}{ccccc}
CHO & CHO & CHO & & CHO \\
\overset{*}{C}HOH & \overset{}{C}OH & H-\overset{*}{C}-OH & & HO-\overset{*}{C}-H \\
CHOH & CHOH & CHOH & + & CHOH \\
CHOH & CHOH & CHOH & & CHOH \\
CHOH & CHOH & CHOH & & CHOH \\
CH_2OH & CH_2OH & CH_2OH & & CH_2OH \\
\text{D-Glucose} & \text{Enolate ion} & \text{D-Glucose} & & \text{D-Mannose}
\end{array}
$$

α–Hydrogen · Loss of stereo centre

Epimers
(Differ in configuration at carbon-2)

(b) When D-glucose is treated with an alkali, it loses α-hydrogen present at C-2 position and forms an enolate ion. This enolate ion (resonance stabilized) undergoes protonation at oxygen to yield an ene–diol intermediate, which results in the formation of an enolate ion in the presence of alkali. On protonation, this enolate ion is rearranged to form D-fructose as shown:

$$
\begin{array}{ccccccc}
\overset{O}{\underset{\displaystyle\overset{|}{C}}{\diagdown}}^{H} & \overset{O}{\underset{\displaystyle\overset{|}{C}}{\diagdown}}^{H} & \overset{O^-}{\underset{\displaystyle\overset{|}{C}}{\diagdown}}^{H} & \overset{HO}{\underset{\displaystyle\overset{|}{C}}{\diagdown}}^{H} & \overset{HO}{\underset{\displaystyle\overset{|}{C}}{\diagdown}}^{H} & \overset{HO}{\underset{\displaystyle\overset{|}{C}}{\diagdown}}^{H} & CH_2OH \\
\end{array}
$$

*CHOH	C—OH	C—OH	C—OH	C—O⁻	C=O	C=O
CHOH ⇌ (⁻OH)	CHOH ↔	CHOH ⇌ (H⁺)	CHOH ⇌ (⁻OH)	CHOH	CHOH ⇌ (H₂O)	CHOH
CHOH	CHOH	CHOH	CHOH	CHOH	CHOH	CHOH
CHOH	CHOH	CHOH	CHOH	CHOH	CHOH	CHOH
CH₂OH	CH₂OH	CH₂OH	CH₂OH	CH₂OH	CH₂OH	CH₂OH
D-Glucose	Enolate ion		ene-diol intermediate	Enolate ion		Fructose

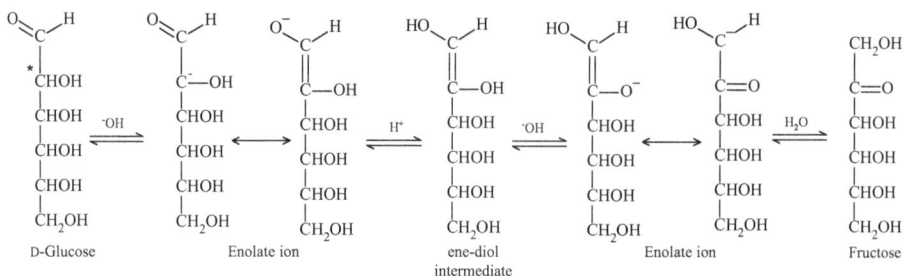

Thus, the interconversion of aldoses and ketoses, that is, glucose to mannose, and glucose to fructose, takes place in an alkaline medium, and proceeds through an ene–diol intermediate. All the steps involved in the reactions are reversible; therefore, interconversion of D-glucose, D-fructose, and D-mannose can be written as follows:

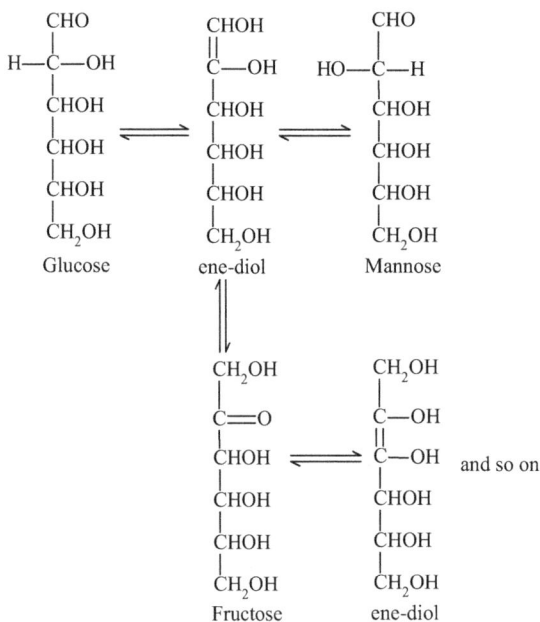

CHO	CHOH	CHO
H—C—OH	C—OH	HO—C—H
CHOH ⇌	CHOH ⇌	CHOH
CHOH	CHOH	CHOH
CHOH	CHOH	CHOH
CH₂OH	CH₂OH	CH₂OH
Glucose	ene-diol	Mannose

⇕

CH₂OH	CH₂OH
C=O	C—OH
CHOH ⇌	C—OH	and so on
CHOH	CHOH
CHOH	CHOH
CH₂OH	CH₂OH
Fructose	ene-diol

1.7 Fructose/Fruit Sugar/Laevulose: Properties and Structure

Fructose is the only naturally occurring ketohexose, also known as fruit sugar, and this has the same molecular formula as that of glucose ($C_6H_{12}O_6$). Fructose was discovered by the French chemist Augustin-Pierre Dubrunfaut in 1847 while the name "fructose" was first coined by English chemist William Allen Miller in 1857, and is derived from Latin *Fructus* meaning fruit and *ose* the generic chemical suffix for sugars. It is present in sweet fruits and honey, and is sweetest of all sugars. Fructose is also called *laevulose* because it is an optically active laevorotatory (–) compound. It is also the main constituent found in inulin, a naturally occurring polysaccharides present in artichokes and dahlias.

1.7.1 Preparation of Fructose

1. From starch
Commercially, fructose is prepared by the hydrolysis of inulin in acidic medium:

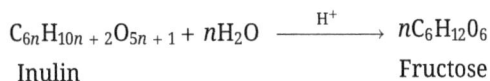

$$C_{6n}H_{10n+2}O_{5n+1} + nH_2O \xrightarrow{\text{H}^+} nC_6H_{12}O_6$$
$$\text{Inulin} \qquad\qquad\qquad\qquad \text{Fructose}$$

2. From sucrose (cane sugar)
Fructose can be prepared in the laboratory by boiling sucrose (cane sugar) with dilute sulfuric acid or with the enzyme invertase for about 2 h. This reaction yields glucose and fructose in equal amount, and fructose can be separated by crystallization:

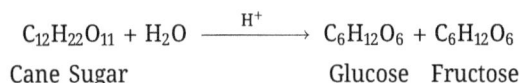

$$C_{12}H_{22}O_{11} + H_2O \xrightarrow{\text{H}^+} C_6H_{12}O_6 + C_6H_{12}O_6$$
$$\text{Cane Sugar} \qquad\qquad \text{Glucose} \quad \text{Fructose}$$

1.7.2 Structure Elucidation of Fructose

Following evidences supported the structure of fructose:

(i) Fructose has the molecular formula $C_6H_{12}O_6$ as shown by elemental analysis and molecular weight determination.

(ii) Fructose on reduction with H_2/Ni forms two epimeric alcohols, sorbitol and mannitol ($C_6H_{14}O_6$), which on treatment with HI leads to the complete reduction of fructose to *n*-hexane. This reaction indicates that six carbon atoms in fructose are present in a straight chain:

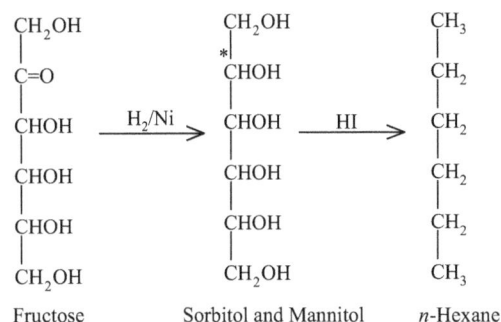

CH$_2$OH		CH$_2$OH		CH$_3$
C=O		*CHOH		CH$_2$
CHOH	H$_2$/Ni	CHOH	HI	CH$_2$
CHOH		CHOH		CH$_2$
CHOH		CHOH		CH$_2$
CH$_2$OH		CH$_2$OH		CH$_3$
Fructose		Sorbitol and Mannitol		*n*-Hexane

(iii) The reaction of fructose with hydroxylamine and phenylhydrazine (1 mol) results in the formation of oxime and fructose phenylhydrazone, respectively. These reactions clearly indicate the presence of a carbonyl group in fructose:

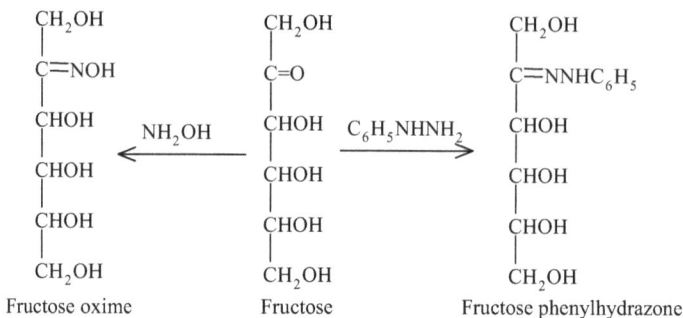

$$
\begin{array}{ccccc}
\text{CH}_2\text{OH} & & \text{CH}_2\text{OH} & & \text{CH}_2\text{OH} \\
| & & | & & | \\
\text{C}{=}\text{NOH} & & \text{C}{=}\text{O} & & \text{C}{=}\text{NNHC}_6\text{H}_5 \\
| & \xleftarrow{\ \text{NH}_2\text{OH}\ } & | & \xrightarrow{\ \text{C}_6\text{H}_5\text{NHNH}_2\ } & | \\
\text{CHOH} & & \text{CHOH} & & \text{CHOH} \\
| & & | & & | \\
\text{CHOH} & & \text{CHOH} & & \text{CHOH} \\
| & & | & & | \\
\text{CHOH} & & \text{CHOH} & & \text{CHOH} \\
| & & | & & | \\
\text{CH}_2\text{OH} & & \text{CH}_2\text{OH} & & \text{CH}_2\text{OH} \\
\text{Fructose oxime} & & \text{Fructose} & & \text{Fructose phenylhydrazone}
\end{array}
$$

(iv) It reacts with HCN to produce a cyanohydrin, which undergoes hydrolysis and reacts with HI giving 2-methylhexanoic acid. This shows that the carbonyl group present in fructose is a keto group (at carbon atom 2):

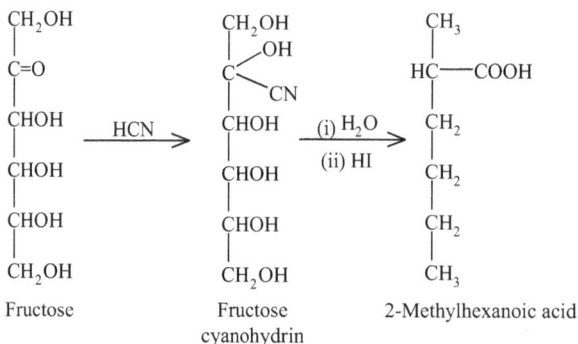

$$
\begin{array}{ccccc}
\text{CH}_2\text{OH} & & \text{CH}_2\text{OH} & & \text{CH}_3 \\
| & & |\ \diagup\text{OH} & & | \\
\text{C}{=}\text{O} & & \text{C}\diagdown & & \text{HC}\!-\!\text{COOH} \\
| & \xrightarrow{\ \text{HCN}\ } & |\ \ \text{CN} & \xrightarrow[\text{(ii) HI}]{\text{(i) H}_2\text{O}} & | \\
\text{CHOH} & & \text{CHOH} & & \text{CH}_2 \\
| & & | & & | \\
\text{CHOH} & & \text{CHOH} & & \text{CH}_2 \\
| & & | & & | \\
\text{CHOH} & & \text{CHOH} & & \text{CH}_2 \\
| & & | & & | \\
\text{CH}_2\text{OH} & & \text{CH}_2\text{OH} & & \text{CH}_3 \\
\text{Fructose} & & \text{Fructose} & & \text{2-Methylhexanoic acid} \\
& & \text{cyanohydrin} & &
\end{array}
$$

(v) On acetylation with acetic anhydride, it forms 1,3,4,5,6-pentaacetate which indicates that it contains five hydroxy groups. The higher stability of the derivative shows that five hydroxy groups must be present on different carbons:

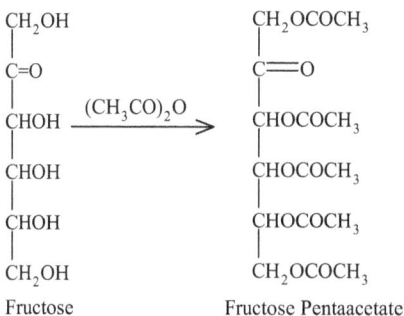

$$
\begin{array}{ccc}
\text{CH}_2\text{OH} & & \text{CH}_2\text{OCOCH}_3 \\
| & & | \\
\text{C}{=}\text{O} & & \text{C}{=}\text{O} \\
| & \xrightarrow{\ (\text{CH}_3\text{CO})_2\text{O}\ } & | \\
\text{CHOH} & & \text{CHOCOCH}_3 \\
| & & | \\
\text{CHOH} & & \text{CHOCOCH}_3 \\
| & & | \\
\text{CHOH} & & \text{CHOCOCH}_3 \\
| & & | \\
\text{CH}_2\text{OH} & & \text{CH}_2\text{OCOCH}_3 \\
\text{Fructose} & & \text{Fructose Pentaacetate}
\end{array}
$$

(vi) Fructose does not react with bromine water:

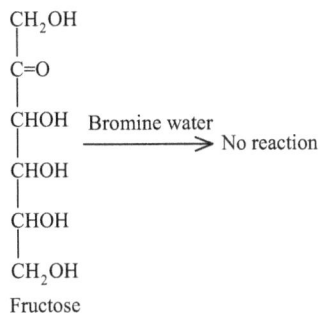

$$
\begin{array}{l}
CH_2OH \\
\;|\; \\
C{=}O \\
\;|\; \\
CHOH \\
\;|\; \\
CHOH \\
\;|\; \\
CHOH \\
\;|\; \\
CH_2OH
\end{array}
\quad
\xrightarrow[\text{Bromine water}]{}
\quad
\text{No reaction}
$$

Fructose

(vii) Fructose on treatment with conc. nitric acid yields a mixture of glutaric acid, tartaric acid, and glycolic acid. These acids contain lesser number of carbon atoms than fructose. This reaction further confirms the presence of a keto group:

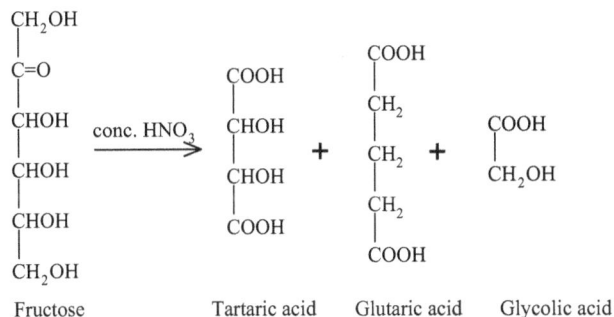

$$
\begin{array}{l}
CH_2OH \\
\;|\; \\
C{=}O \\
\;|\; \\
CHOH \\
\;|\; \\
CHOH \\
\;|\; \\
CHOH \\
\;|\; \\
CH_2OH
\end{array}
\;\;\xrightarrow{\text{conc. } HNO_3}\;\;
\begin{array}{l}
COOH \\
\;|\; \\
CHOH \\
\;|\; \\
CHOH \\
\;|\; \\
COOH
\end{array}
\;+\;
\begin{array}{l}
COOH \\
\;|\; \\
CH_2 \\
\;|\; \\
CH_2 \\
\;|\; \\
CH_2 \\
\;|\; \\
COOH
\end{array}
\;+\;
\begin{array}{l}
COOH \\
\;|\; \\
CH_2OH
\end{array}
$$

Fructose   Tartaric acid   Glutaric acid   Glycolic acid

(viii) Glucose and fructose on treatment with phenylhydrazine produce the same kind of osazone, and since the configuration of glucose had already been established, it helped in determining the configuration of D-fructose.

On the basis of abovementioned reactions, it was established that fructose has a six-carbon straight chain and a keto group at carbon atom 2. Thus, the structure of fructose can be represented as follows:

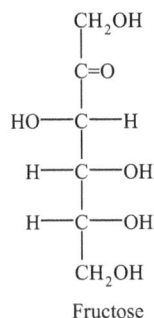

$$
\begin{array}{l}
CH_2OH \\
\;|\; \\
C{=}O \\
\;|\; \\
HO{-}C{-}H \\
\;|\; \\
H{-}C{-}OH \\
\;|\; \\
H{-}C{-}OH \\
\;|\; \\
CH_2OH
\end{array}
$$

Fructose

The various reactions of fructose are summarized in Figure 1.11.

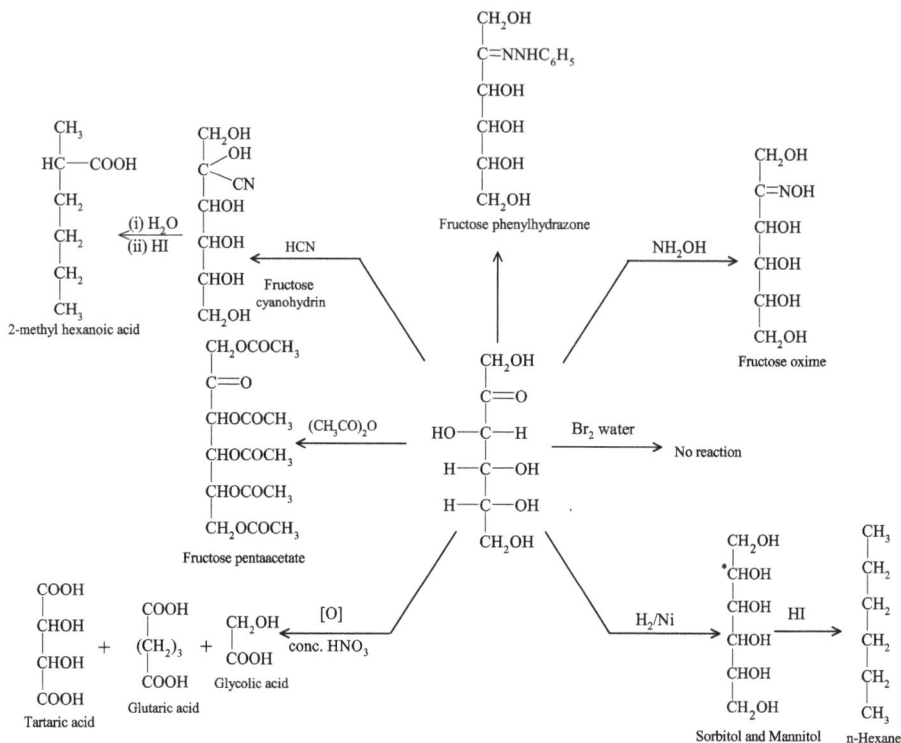

Figure 1.11: Reactions of fructose.

1.7.3 Osazone Formation: Reaction of Fructose with Excess of Phenylhydrazine

Fructose when treated with excess of phenylhydrazine yields fructose osazone:

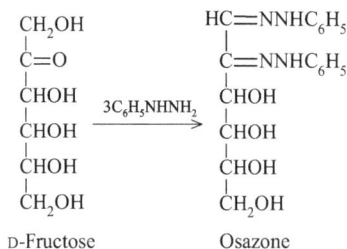

In the first step, carbonyl (keto) group of fructose reacts with phenylhydrazine to form fructose phenylhydrazone. In the second step, the Amadori rearrangement takes place, in which CH_2OH group (at carbon atom 1) is oxidized to aldehyde group (-CHO), followed by the elimination of aniline. The third step involves the reaction between newly generated carbonyl group with the second mole of phenylhydrazine, followed by reaction with the third mole of phenylhydrazine resulting in the formation of fructose osazone. Figure 1.12 displays the steps involved in the formation of fructose osazone.

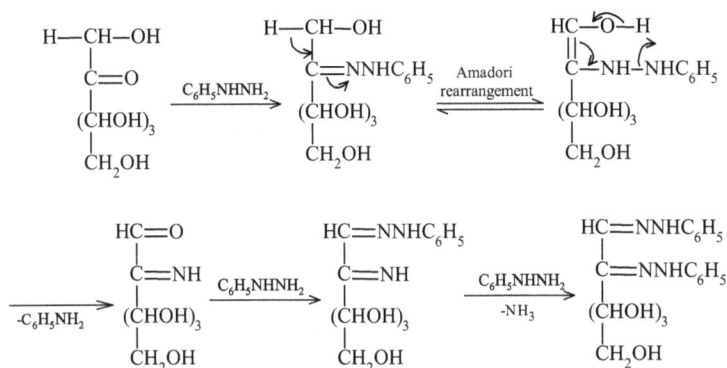

Figure 1.12: Mechanism of osazone formation.

1.7.4 Cyclic Structure of D-Fructose

The open chain structure of fructose does not account for every reaction; therefore, like glucose, fructose also exists in cyclic form. Here are few reactions which indicate toward cyclic structure of fructose:

(i)   Though a carbonyl group is present in fructose, it does not react with sodium bisulfite.

(ii) Mutarotation is also observed in fructose, and it exists in two isomeric forms, namely α-D-fructose and β-D-fructose, which is possible only when fructose exists in cyclic forms.

(iii) Two methyl fructosides, that is, methyl-α-D-fructoside and methyl-β-D-fructoside also exist which again confirms the cyclic structure of fructose as well as two isomeric forms, α and β.

In cyclic form of fructose, the intramolecular cyclization occurs between carbon atom 2 (C-2) and carbon atom 6 (C-6), and it exists as six-membered rings, as α-D-fructopyranose and β-D-fructopyranose. Fructose occurs in pyranose form in free state; however, it exists as α-D-fructofuranose and β-D-fructofuranose in the combined state. The cyclic structures of α-D-fructose and β-D-fructose are shown in Figure 1.13.

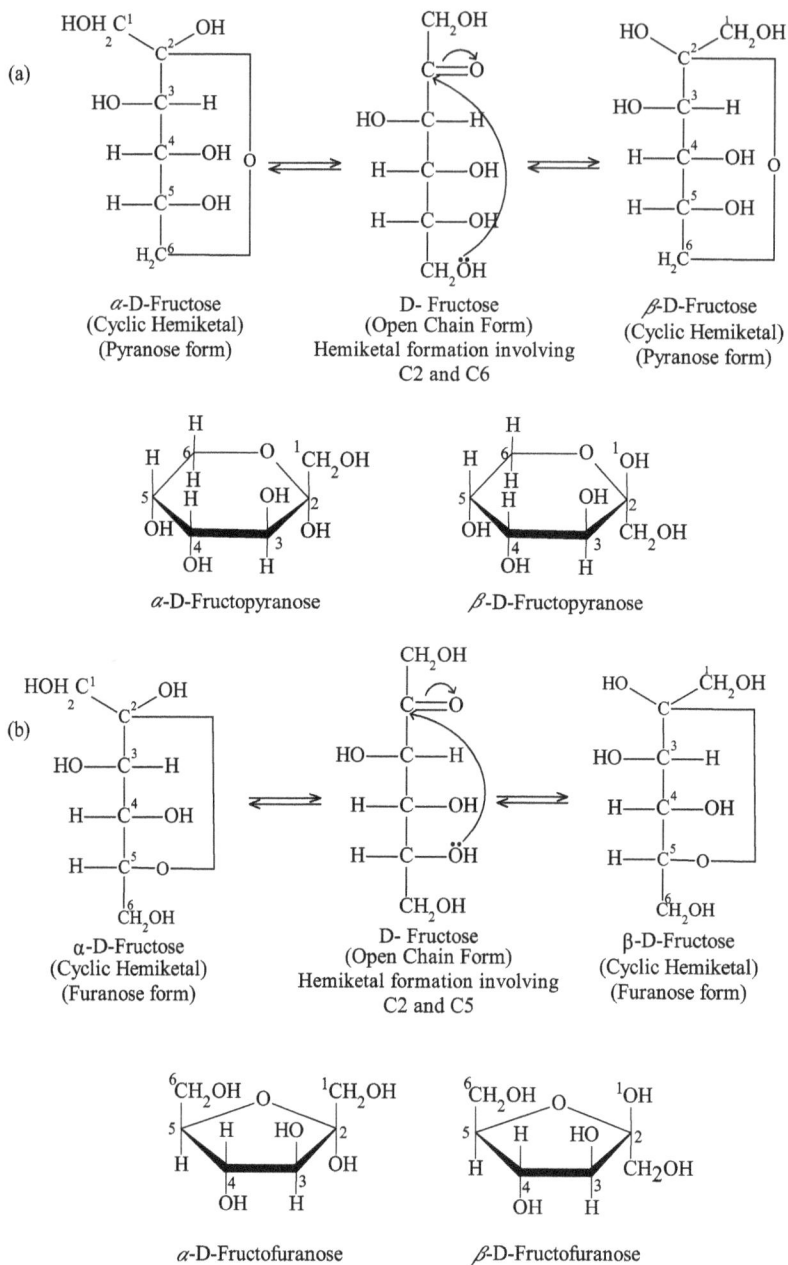

(a)

α-D-Fructose
(Cyclic Hemiketal)
(Pyranose form)

D- Fructose
(Open Chain Form)
Hemiketal formation involving
C2 and C6

β-D-Fructose
(Cyclic Hemiketal)
(Pyranose form)

α-D-Fructopyranose

β-D-Fructopyranose

(b)

α-D-Fructose
(Cyclic Hemiketal)
(Furanose form)

D- Fructose
(Open Chain Form)
Hemiketal formation involving
C2 and C5

β-D-Fructose
(Cyclic Hemiketal)
(Furanose form)

α-D-Fructofuranose

β-D-Fructofuranose

Figure 1.13: Cyclic structures of D-fructose: (a) six-membered cyclic hemiketal form and (b) five-membered cyclic hemiketal form.

1.8 Important Conversions of Monosaccharides

1.8.1 D-Glucose to D-Fructose (Aldose to Ketose)

Glucose (an aldose) on treatment with 3 mol of phenylhydrazine gives glucosazone, which undergoes hydrolysis to yield osone. On reduction with sodium amalgam, osone is converted into fructose (a ketose):

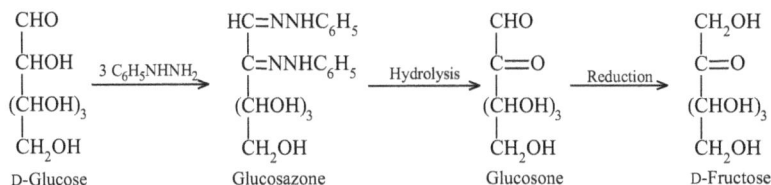

$$
\begin{array}{ccccc}
\text{CHO} & & \text{HC=NNHC}_6\text{H}_5 & & \text{CHO} & & \text{CH}_2\text{OH} \\
\text{CHOH} & \xrightarrow{3\ \text{C}_6\text{H}_5\text{NHNH}_2} & \text{C=NNHC}_6\text{H}_5 & \xrightarrow{\text{Hydrolysis}} & \text{C=O} & \xrightarrow{\text{Reduction}} & \text{C=O} \\
\text{(CHOH)}_3 & & \text{(CHOH)}_3 & & \text{(CHOH)}_3 & & \text{(CHOH)}_3 \\
\text{CH}_2\text{OH} & & \text{CH}_2\text{OH} & & \text{CH}_2\text{OH} & & \text{CH}_2\text{OH} \\
\text{D-Glucose} & & \text{Glucosazone} & & \text{Glucosone} & & \text{D-Fructose}
\end{array}
$$

1.8.2 D-Fructose to D-Glucose (Ketose to Aldose)

On catalytic reduction, fructose yields a mixture of hexahydric alcohols, that is, sorbitol and mannitol. Further, hexahydric alcohol on oxidation produces a monocarboxylic acid, which on heating yields lactone. Lactone, when reduced with sodium amalgam, forms D-glucose:

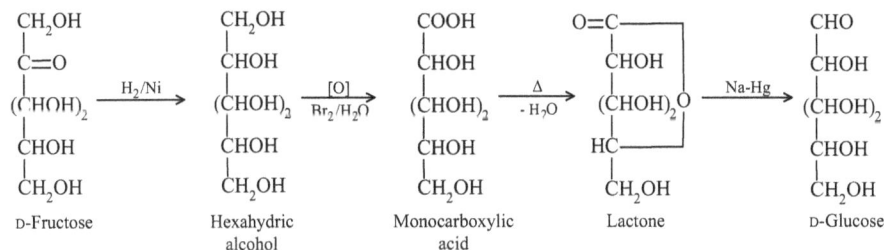

$$
\begin{array}{ccccccccc}
\text{CH}_2\text{OH} & & \text{CH}_2\text{OH} & & \text{COOH} & & \text{O=C}\!\!-\!\! & & \text{CHO} \\
\text{C=O} & & \text{CHOH} & & \text{CHOH} & & \text{CHOH} & & \text{CHOH} \\
\text{(CHOH)}_2 & \xrightarrow{\text{H}_2/\text{Ni}} & \text{(CHOH)}_2 & \xrightarrow[\text{Br}_2/\text{H}_2\text{O}]{\text{[O]}} & \text{(CHOH)}_2 & \xrightarrow[-\,\text{H}_2\text{O}]{\Delta} & \text{(CHOH)}_2\,\text{O} & \xrightarrow{\text{Na-Hg}} & \text{(CHOH)}_2 \\
\text{CHOH} & & \text{CHOH} & & \text{CHOH} & & \text{HC}\!\!-\!\! & & \text{CHOH} \\
\text{CH}_2\text{OH} & & \text{CH}_2\text{OH} & & \text{CH}_2\text{OH} & & \text{CH}_2\text{OH} & & \text{CH}_2\text{OH} \\
\text{D-Fructose} & & \text{Hexahydric} & & \text{Monocarboxylic} & & \text{Lactone} & & \text{D-Glucose} \\
& & \text{alcohol} & & \text{acid}
\end{array}
$$

1.8.3 D-Glucose to D-Mannose (Aldose to Its Epimer)

D-Glucose forms gluconic acid on oxidation with bromine water, which on further reaction with pyridine forms epimeric aldonic acid, gluconic acid. Gluconic acid is then separated and undergoes dehydration to yield a lactone. The resulting lactone on reduction with sodium amalgam yields D-mannose.

```
  H    O                                                                              H    O
   \ //          COOH              COOH           O=C                                  \ //
    C             |                 |             |   ‾‾‾‾‾‾‾|                           C
    |             |                 |          HO—C—H       |                           |
 H—C—OH        H—C—OH           HO—C—H          |           |                        HO—C—H
    |   Br₂ water  |     Pyridine    |    -H₂O   CHOH    O   |     Na-Hg                 |
  CHOH  ──────→  CHOH   ──────→    CHOH   ──────→ |           |  ──────→              CHOH
    |             |                 |          CHOH         |                           |
  CHOH          CHOH              CHOH           |           |                        CHOH
    |             |                 |           HC ‾‾‾‾‾‾‾‾‾‾                            |
  CHOH          CHOH              CHOH           |                                    CHOH
    |             |                 |          CH₂OH                                    |
  CH₂OH         CH₂OH             CH₂OH                                               CH₂OH
 Glucose      Gluconic acid                                                          Mannose
```

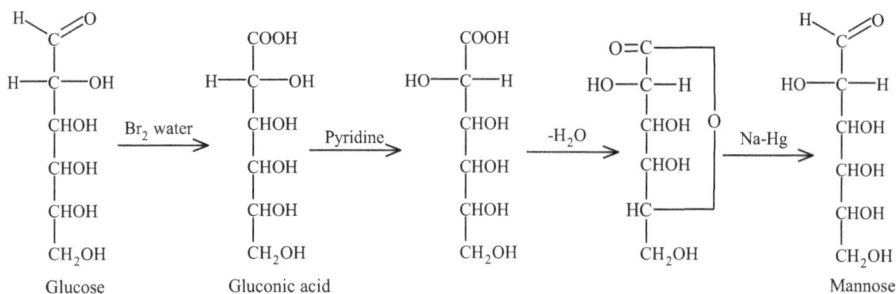

1.9 Disaccharides

Disaccharides are sugars comprising two monosaccharide units, which may be same or different, linked by an O-glycosidic bond. A disaccharide can be formed from the reaction of the anomeric carbon of one cyclic monosaccharide with the -OH group of a second monosaccharide. The most important disaccharides are sucrose, maltose, and lactose. Their general formula is $C_{12}H_{22}O_{11}$, but they differ in their monosaccharide constituents as well as the specific type of glycosidic linkage connecting them. Disaccharides, on hydrolysis with dilute acids or enzymes, yield two molecules of monosaccharides:

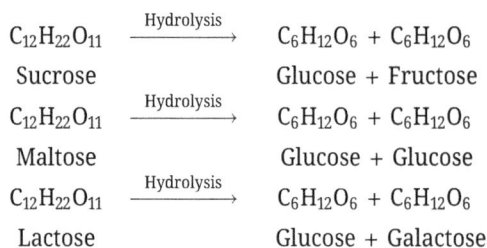

$$C_{12}H_{22}O_{11} \xrightarrow{\text{Hydrolysis}} C_6H_{12}O_6 + C_6H_{12}O_6$$

Sucrose $\qquad\qquad$ Glucose + Fructose

$$C_{12}H_{22}O_{11} \xrightarrow{\text{Hydrolysis}} C_6H_{12}O_6 + C_6H_{12}O_6$$

Maltose $\qquad\qquad$ Glucose + Glucose

$$C_{12}H_{22}O_{11} \xrightarrow{\text{Hydrolysis}} C_6H_{12}O_6 + C_6H_{12}O_6$$

Lactose $\qquad\qquad$ Glucose + Galactose

1.9.1 Sucrose (Cane Sugar)

Sucrose, also known as table sugar, is the most common disaccharide found in nature. Commercially, it is obtained from sugar cane and sugar beet. Sucrose is also found in varying amounts in many fruits such as pineapple, ripe banana, and ripe mangoes. Sucrose has the molecular formula $C_{12}H_{22}O_{11}$, and is composed of D-glucose and D-fructose.

Sucrose is a colorless, crystalline substance and sweet in nature. It is easily soluble in water and is dextrorotatory having specific rotation, $[\alpha]_D = +66.5°$. Sucrose on hydrolysis in dilute acidic medium or through the action of the enzyme sucrase yields equal amounts of glucose and fructose, known as invert sugar. The hydrolysis reaction involves a change in specific rotation sign from positive to negative; it is therefore termed as inversion of (+) sucrose, and the resulting 1:1 mixture of D-glucose and D-fructose ($[\alpha]_D = -20°$) is called invert sugar:

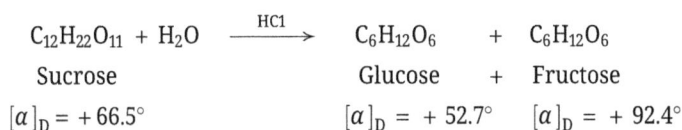

$$\underset{\substack{\text{Sucrose} \\ [\alpha]_D = +66.5°}}{C_{12}H_{22}O_{11}} + H_2O \xrightarrow{\text{HC1}} \underset{\substack{\text{Glucose} \\ [\alpha]_D = +52.7°}}{C_6H_{12}O_6} + \underset{\substack{\text{Fructose} \\ [\alpha]_D = +92.4°}}{C_6H_{12}O_6}$$

It is used as a sweetening agent in various food preparations such as jams, syrups, sweets, as well as food preservatives. Sucrose octaacetate is specifically used in manufacturing of transparent papers and in preparation of nonaqueous adhesives.

1.9.1.1 Structure Elucidation of Sucrose

The structure of sucrose is based on the following observations:

(i) Elemental analysis shows that the molecular formula of sucrose is $C_{12}H_{22}O_{11}$.

(ii) It yields an octaacetate and an octamethyl derivative, which indicate the presence of eight hydroxy groups.

(iii) Sucrose is a nonreducing sugar. It gives a negative test with Fehling's solution and does not reduce Tollens' reagent. Moreover, it does not form an osazone, does not show mutarotation in solution, and does not exist in anomeric forms. All these facts reveal that sucrose does not contain a free aldehyde or ketonic group in its structure. It also does not react with hydroxylamine and phenylhydrazine.

(iv) Acid hydrolysis of sucrose produces a mixture of D-glucose and D-fructose. Since there is no "free" carbonyl group, it is possible that glucose and fructose must have been linked through C-1 carbon of glucose (carrying -CHO group) and C-2 carbon of fructose (carrying a ketonic group) as only in this way both carbonyl functions can be effectively blocked by a single link.

(v) The ring size of D-glucose and D-fructose in sucrose is established on the basis of usual methylation studies. Sucrose, on methylation, gives an octamethyl derivative, which on hydrolysis produces 2,3,4,6-tetra-*O*-methyl glucose and 1,3,4,6-tetra -*O*-methyl fructose. Formation of 2,3,4,6-tetra-*O*-methyl glucose indicates the presence of ring formation between C-1 and C-5 of glucose unit, whereas formation of 1,3,4,6-tetra-*O*-methyl derivative of fructose depicts the presence of a ring formation between C-2 and C-5 of fructose unit. So, it was confirmed that in sucrose, glucose is in the pyranose form and fructose is in the furanose form.

(vi) The configuration at the glycosidic linkage was established by enzymatic studies. Sucrose is hydrolyzed by an α-glucosidase, and not by β-glucosidase, which indicates α-configuration at the glucoside portion of the molecule. The hydrolysis of sucrose is catalyzed by sucrase, an enzyme that specifically hydrolyzes β-fructofuranosides, which shows the presence of β-configuration at fructose.

Thus, it has been proposed that sucrose consists of α-D-glucopyranosyl-β-D-fructofuranoside, that is, α-glucose is linked to β-fructose.

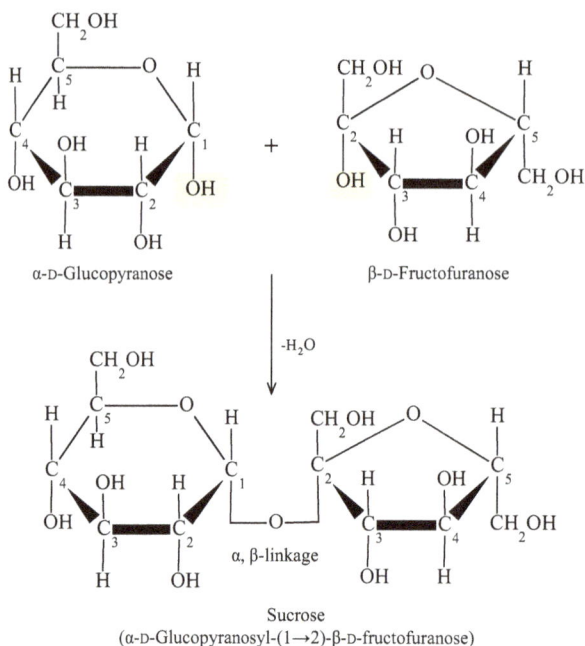

α-D-Glucopyranose   +   β-D-Fructofuranose

-H$_2$O

Sucrose
(α-D-Glucopyranosyl-(1→2)-β-D-fructofuranose)

1.9.2 Lactose (Milk Sugar)

Lactose is found solely in milk of mammals, for example, cow's milk contains about 5% and human milk contains about 7% lactose. Lactose has the molecular formula $C_{12}H_{22}O_{11}$ and is a reducing sugar consisting of one molecule of D-glucose and one molecule of D-galactose. Milk sours when lactose is converted into lactic acid by the action of bacteria, *Lactobacillus bulgaricus*. Commercially, lactose is prepared from whey, a by-product of cheese and casein production, crystallizing an oversaturated solution of whey concentrate to get lactose.

It is a white crystalline solid, easily soluble in water, whereas insoluble in other solvents such as alcohol, ether, and benzene. It forms an osazone and exists in α- and β-forms which undergo mutarotation and exists in equilibrium.

1.9.2.1 Structure Elucidation of Lactose
The structure of lactose is established on the basis of following facts:
(i)   Hydrolysis of lactose produces 1 mol of glucose and 1 mol of galactose.
(ii) Lactose is a reducing disaccharide. It reduces Fehling's solution as well as Tollens' reagent. It also reacts with HCN and forms osazone.
(iii) It exists in two isomeric forms, α and β, which undergo mutarotation:

	α-form	⇌	Equilibrium mixture	⇌	β-form
Specific rotation	+ 90°		+ 55.4°		+ 35°

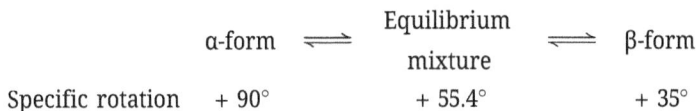

(iv) Lactobionic acid ($C_{12}H_{22}O_{12}$), a monocarboxylic acid, is obtained by the chemical oxidation of lactose. On hydrolysis, lactobionic acid gives a mixture of D-galactose and D-gluconic acid. This reaction confirms that the galactose moiety is in the non-reducing part of the molecule, and it is the D-glucose unit that has a "free" aldehyde group and shows osazone formation and oxidation:

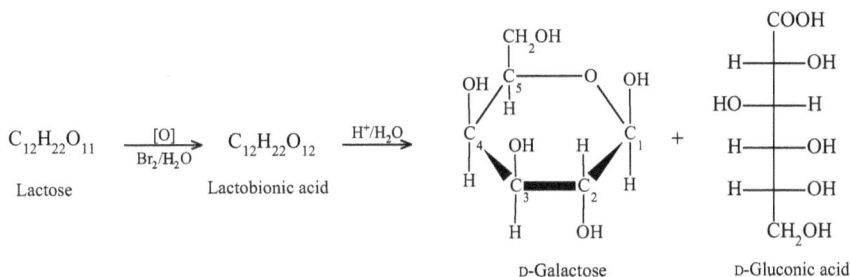

$$C_{12}H_{22}O_{11} \xrightarrow[Br_2/H_2O]{[O]} C_{12}H_{22}O_{12} \xrightarrow{H^+/H_2O}$$

Lactose   Lactobionic acid

D-Galactose   D-Gluconic acid

(v) The glycosidic linkage between glucose and galactose in lactose is established using methylation method. Lactose undergoes methylation with dimethyl sulfate in alkaline medium, and yields an octa-O-methyl lactose, which on hydrolysis produces a mixture of 2,3,4,6-tetra-O-methyl-D-galactopyranose and 2,3,6-tri-O-methyl-D-glucopyranose. This indicates the presence of six-membered pyranose ring in both the units and shows that C-4 hydroxy group of the glucose is bound in the glycosidic linkage.

(vi) Further, on complete methylation, lactobionic acid gives methyl-octa-O-methyl lactobionate, which on acid hydrolysis produces 2,3,4,6-tetra-O-methyl-D-galactopyranose and 2,3,5,6-tetra-O-methyl-D-gluconic acid. This study shows that anomeric hydroxy (C-1) of the galactose is bound to C-4 of the glucose moiety through glycosidic linkage.

(vii) The hydrolysis of lactose is catalyzed by β-galactosides, an enzyme specifically catalyze β-galactoside linkages:

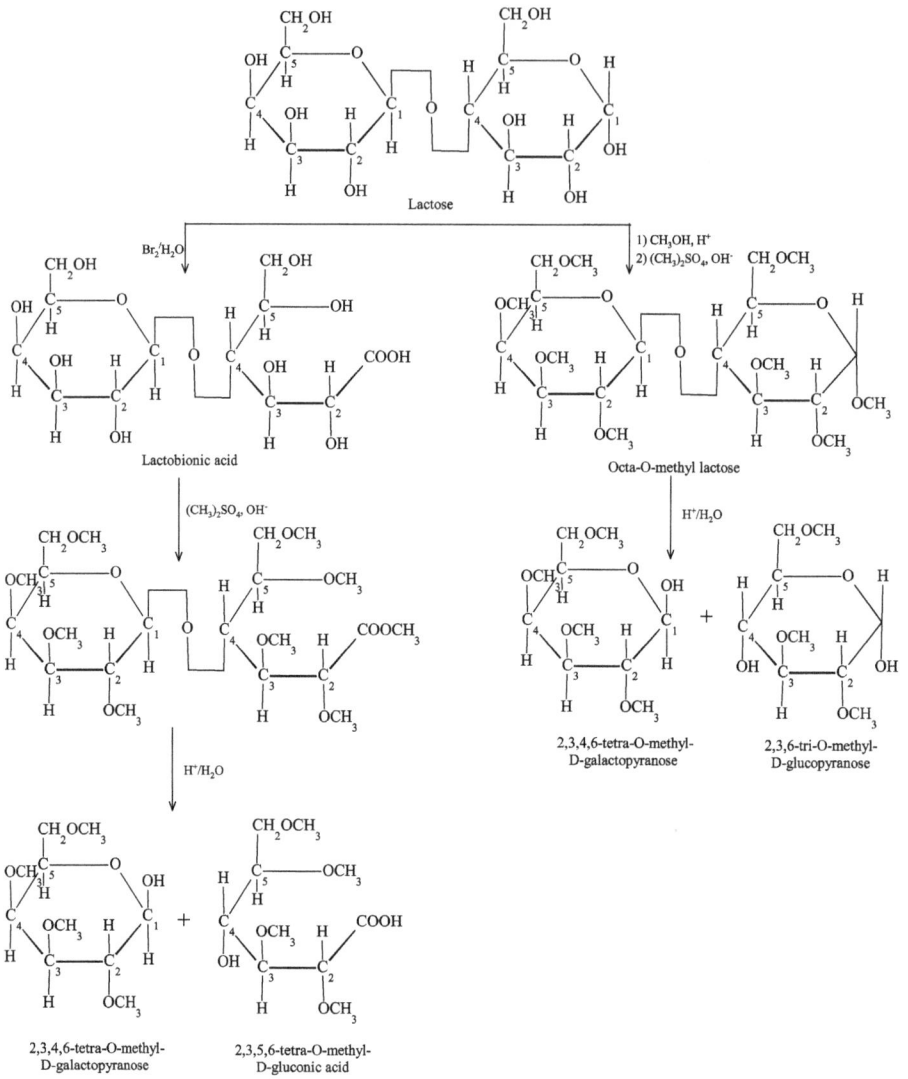

Lactose

Lactobionic acid

Octa-O-methyl lactose

2,3,4,6-tetra-O-methyl-
D-galactopyranose

2,3,6-tri-O-methyl-
D-glucopyranose

2,3,4,6-tetra-O-methyl-
D-galactopyranose

2,3,5,6-tetra-O-methyl-
D-gluconic acid

Based on all the abovementioned facts, lactose is designated as β-D-galactopyranosyl
-(1→4)-α-D-glucopyranose, and its structure is

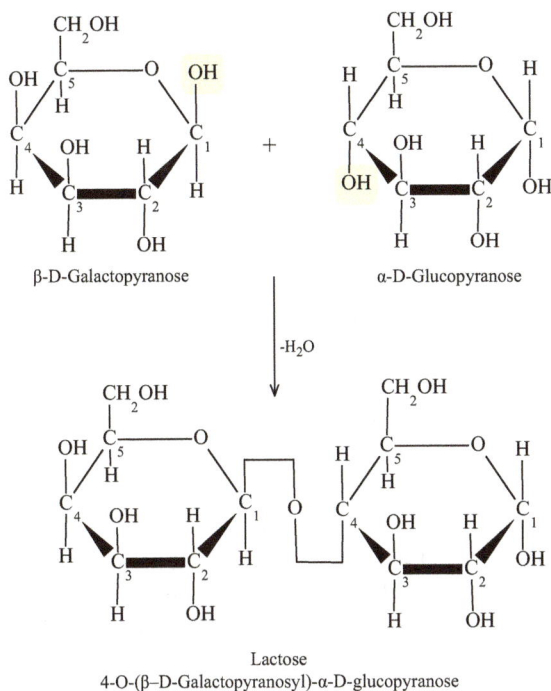

β-D-Galactopyranose   α-D-Glucopyranose

Lactose
4-O-(β–D-Galactopyranosyl)-α-D-glucopyranose

1.9.3 Maltose (Malt Sugar)

Maltose can be obtained by the partial hydrolysis of starch in aqueous solution, which in turn hydrolyzed to two units of glucose by maltase. It is also known as malt sugar. In maltose, two glucose units are linked together by an α-1,4-glycosidic linkage. It exists in both α- and β-forms, and reduces Tollens' as well as Fehling's solution.

1.9.3.1 Structure Elucidation of Maltose
The following facts establish the structure of maltose:
(i)   Maltose yields two molecules of D-(+)-glucose, when hydrolyzed in aqueous acid or treated with maltase.
(ii) Maltose has the molecular formula, $C_{12}H_{22}O_{11}$. It is a reducing sugar, that is, it gives positive test with Fehling's solution, Tollens' reagent, and Benedict reagent.
(iii) Maltose contains only "one" free carbonyl group.
(iv) It has a carbonyl group that exists in the reactive hemiacetal form similar to monosaccharides.
(v)   It also reacts with three molecules of phenylhydrazine to form a phenylosazone.

(vi) When treated with bromine water, it gets oxidized to a monocarboxylic acid (maltobionic acid, $(C_{11}H_{21}O_{10})COOH$).

(vii) Maltose exists in two anomeric forms and shows mutarotation:

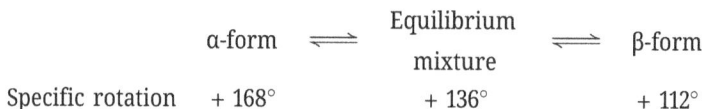

	α-form	⇌	Equilibrium mixture	⇌	β-form
Specific rotation	+ 168°		+ 136°		+ 112°

This contemplation clearly reveals that one of the glucose units in maltose exists in a hemiacetal form, whereas the other is present as a glucoside.

(viii) On methylation, maltose yields an octa-*O*-methyl derivative, which is nonreducing. Further, this octa-methyl derivative on hydrolysis produces a mixture of 2,3,4,6-tetra-*O*-methyl-D-glucopyranose, and 2,3,6-tri-*O*-methyl-D-glucopyranose. This indicates that free -OH at C-5 in the second product is involved in the oxide ring.

(ix) Oxidation of maltose with bromine water gives a monocarboxylic acid (D-maltobionic acid). Methylation of maltobionic acid with methylsulfate and sodium hydroxide yields octa-*O*-methyl maltobionic acid, which under acidic hydrolysis produces 2,3,4,6-tetra-*O*-methyl-D-glucopyranose and 2,3,5,6-tetra-*O*-methyl-D-gluconic acid. This indicates that second product has free hydroxy at C-4 that must have been involved in the glycosidic linkage.

(x) On the basis of abovementioned facts, it can be inferred that both the glucose units in maltose are in the pyranose form, and C-4 hydroxyl of the reducing sugar is involved in the glycosidic bond formation with C-1 hydroxyl of the non-reducing sugar.

(xi) The stereochemistry of the glycosidic linkage in maltose is established on the basis of enzymatic hydrolysis. Maltose is hydrolyzed by maltase, an α-glycosidase enzyme that specifically catalyzed α-glycosidic linkages, which confirmed that the glycosidic linkage in maltose is α.

CH_2OH CH_2OH

Maltose

Br_2/H_2O

1) CH_3OH, H^+
2) $(CH_3)_2SO_4$, OH^-

CH_2OH CH_2OH

Maltobionic acid

CH_2OCH_3 CH_2OCH_3

Octa-O-methyllactose

$(CH_3)_2SO_4$, OH^-

H^+/H_2O

CH_2OCH_3 CH_2OCH_3

CH_2OCH_3 CH_2OCH_3

2,3,4,6-tetra-O-methyl-
D-glucopyranose

2,3,6-tri-O-methyl-
D-glucopyranose

H^+/H_2O

CH_2OCH_3 CH_2OCH_3

$COOH$

$COOCH_3$

2,3,4,6-tetra-O-methyl-
D-glucose

2,3,5,6-tetra-O-methyl-
D-gluconic acid

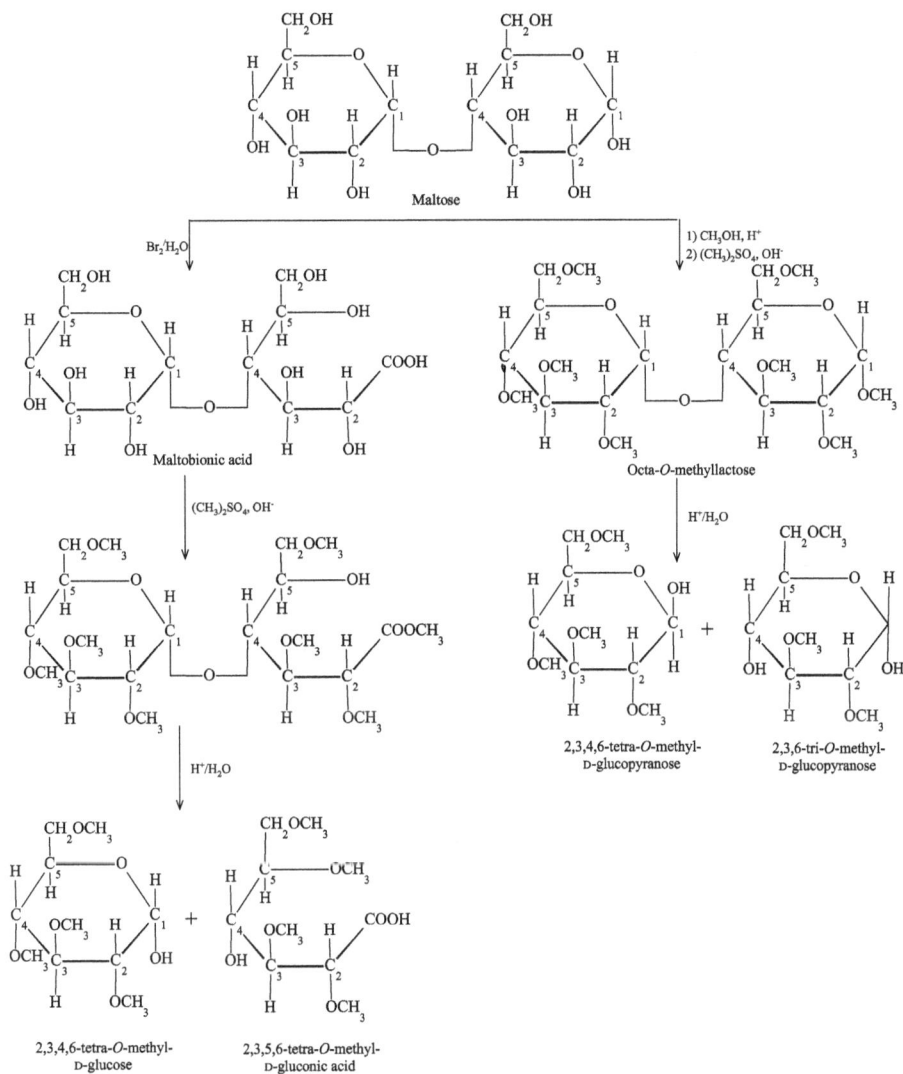

Therefore, on the basis of all these reactions, maltose is designated as α-D-glucopyranosyl -(1→4)-α-D-glucopyranose, and its structure is

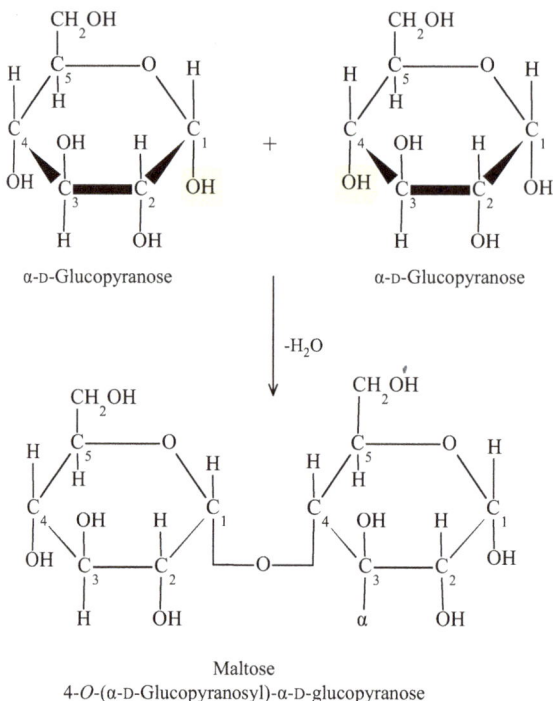

α-ᴅ-Glucopyranose   +   α-ᴅ-Glucopyranose

-H₂O

Maltose
4-*O*-(α-ᴅ-Glucopyranosyl)-α-ᴅ-glucopyranose

1.10 Polysaccharides

Polysaccharides are ubiquitous high-molecular-weight biopolymers that occur widely in nature. These are formed by the association of large number of monosaccharides linked through the glycosidic bond (-O-), and even some of the largest polysaccharide structures are made up of 10,000 individual monomer units. Polysaccharides can be categorized into homopolysaccharides (made up of the same type of monosaccharides) and heteropolysaccharides (made up of different monosaccharides). Different types of polysaccharides exist in nature that differ in the molecular structure (can be linear or branched), type of monosaccharide, linkage between the monosaccharides, and complexity of the overall molecule. Cellulose, starch, glycogen, amylopectin, xylan, amylose, and chitosan are some of the examples of polysaccharides. Glycogen, starch, amylose, and amylopectin are used for storing sugars in plants and animals, whereas cellulose, chitosan, and xylan are used as a structural material to construct cell walls of plants, insects, and crustaceans. The three most common and abundant polysaccharides are starch, glycogen, and cellulose.

1.10.1 Starch

Starch is one of the most important polysaccharides which is found in the stem, roots, tuber, and seeds of plants. It consists of numerous D-glucose units joined by glycosidic linkage. This is the most common carbohydrate in human diets across the world and is present in large amounts in potatoes, wheat, maize, and barley. Starch is a nonreducing sugar as it does not reduce Fehling's solution and Tollen's reagent, and also it does not form osazone. Starch, when heated at higher temperature, decomposes into dextrins and other compounds. Dextrins are a group of low-molecular-weight carbohydrates which are obtained by the hydrolysis of starch and glycogen. The complete hydrolysis of starch yields glucose:

$$starch \rightarrow dextrins \rightarrow maltose \rightarrow glucose$$

Starch granules are semicrystalline, that is, they contain both crystalline and amorphous parts. It is tasteless and odorless white powder that is insoluble in cold water and alcohol. Starch is a mixture of two glucose polymers; it consists of about 20% water-soluble fraction called amylose and 80% of a water-insoluble fraction called amylopectin. Amylose is a linear polysaccharide in which D-glucose residues are linked by the α-1,4-glycosidic linkages. Experimental data indicate that amylose is not a straight chain of glucose units but tend to adopt a helical structure with six glucose monomers per turn. Molecular weight of amylose ranges between 10,000 and 500,000, and it contains 100–3,000 D-glucose units.

Amylopectin is also composed of glucose units linked primarily by α-1,4-glycosidic bonds but it may also contain a few α-1,6 branching points, occurring at an interval of 20–25 glucose units. Amylopectin has high molecular weight, that is, it ranges from 50,000 to 100,000 having 300–6,000 D-glucose units. Starch gives an intense blue color with iodine due to the formation of the amylose–iodine complex, and this color test can detect even minute amounts of starch in solution. The structure of amylose and amylopectin is shown in Figure 1.14.

1.10.2 Glycogen

Glycogen is an important glucose polysaccharide that is found in the liver and muscle tissues of animals as well as in yeast and mushrooms. Structurally, it is quite similar to amylopectin, although glycogen is more highly branched, and branching occurs at every 8–12 glucose units. Branching enhances its solubility, and the rate at which glucose can be stored and retrieved. Glycogen can be broken down into its D-glucose subunits on acidic hydrolysis, or by the action of enzymes. It plays a crucial role as a cellular energy source and serves as an energy store. The excessive glucose is stored as glycogen, and the stored glycogen, when needed, gets converted to glucose. It is an

(a) Amylose

(b) Amylopectin

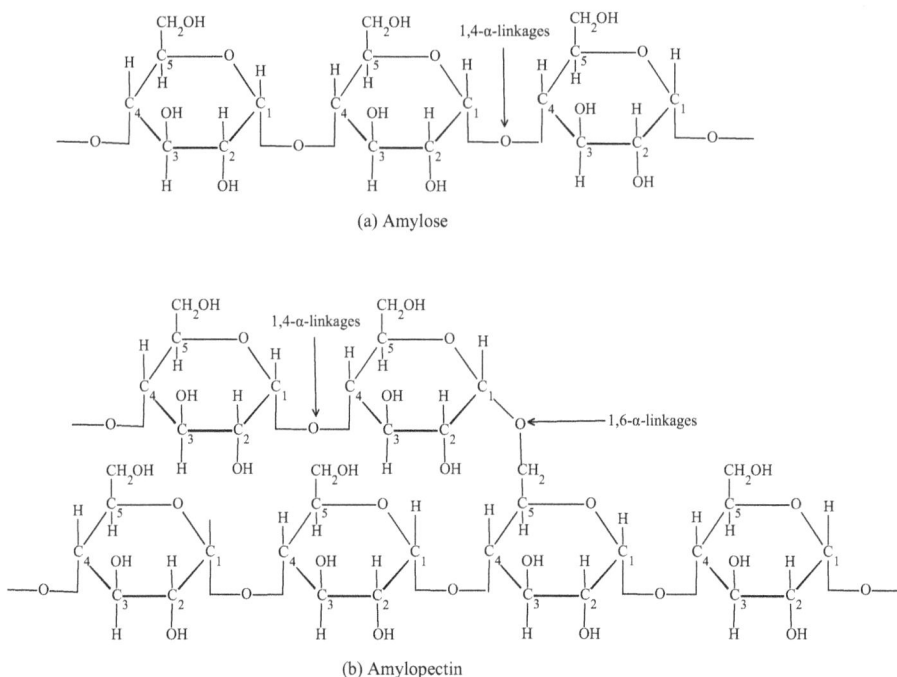

Figure 1.14: Structure of amylose and amylopectin, two constituents of starch.

amorphous powder, easily soluble in water, and produces reddish brown color on re-action with iodine solution. Glycogen is also called animal starch.

1.10.3 Cellulose

Cellulose, the major structural component of the plant cell walls, is the most copious polysaccharide found in nature. It is the major structural component of the plant cell walls. Jute, cotton, hemp, and flax are some of the natural sources of cellulose; other sources are straw, corncobs, and bagasse. It is an odorless, tasteless, amorphous white solid, insoluble in water and most organic solvents. Cellulose has the formula $(C_6H_{10}O_5)_n$, and on complete hydrolysis yields only D-glucose. Structurally, it is similar to starch but in cellulose, D-glucose units are linked together via 1,4-β-glycosidic linkages (Figure 1.15). The molecular weight for cellulose ranges from 250,000 to 1,000,000 or more; that is, it consists of at least 1,500 glucose units per molecule.

Cellulose is the major constituent of paper and paper products. It is the main in-gredient of textiles; filter paper and cotton fibrils are almost completely cellulose, and wood is about 50% cellulose. Cellulose for industrial applications is chiefly obtained from wood pulp and cotton.

The complete hydrolysis of cellulose with dilute acids yields D-glucose; however, cellobiose, a disaccharide, is obtained on incomplete hydrolysis of cellulose. It is important to mention that all the ruminants such as cattle and goat have symbiotic bacteria in the intestinal tract, and these bacteria produce necessary enzymes called *cellulases* that can break down cellulose to glucose. These animals can feed on cellulose directly, whereas humans and many other animals lack enzymes needed to break down cellulose.

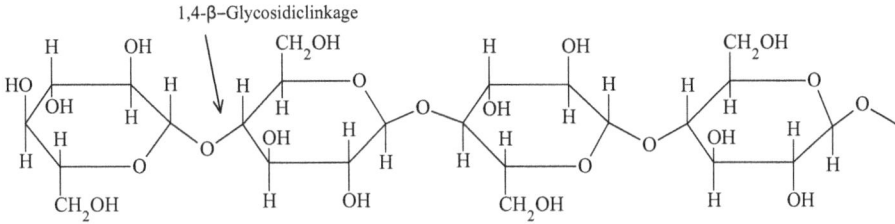

Figure 1.15: Structure of cellulose, a linear polymer of glucose with 1,4-β-linkages.

Lipids

Lipids constitute another class of naturally occurring molecules which are soluble in non-polar organic solvents such as acetone, benzene, and chloroform, but almost insoluble in water. The word "lipids" come from the Greek word *lipos*, which means "fat." Lipids comprised heterogeneous group of biomolecules that are esters of fatty acids like fats, oils, waxes, sterols, fat-soluble vitamins (such as A, D, E, and K), triglycerides, and phospholipids as a result of their physical properties rather than their structures (Figure 1.16). Lipids contain the same elements as carbohydrates: carbon, hydrogen, and oxygen but the hydrogen-to-oxygen ratio is always greater than 2:1. The hydrocarbon portion of the compound is responsible for the "oiliness" or "fattiness" associated with lipids. They are synthesized in our body using various complex biosynthetic pathways; however, similar to amino acids, some lipids are essential and need to be taken in our diet.

Lipids perform numerous essential biological functions in the body; for instance, they act as chemical messengers and signal molecules, form the structural components of cell membranes, and are used for storing energy, local hormonal regulation, vitamin absorption, and many other body functions. Fat serves as an efficient source of energy, both directly and potentially when stored in adipose tissue. They also form a protective coating on many organisms, and protect and insulate them from extreme environmental conditions. The fat content of nerve tissue is particularly high. The fats combined with proteins to form lipoproteins that serve as important cellular constituents in the cell membrane and in the mitochondria within the cytoplasm. Lipoproteins also work as vehicles for transport of lipids in blood.

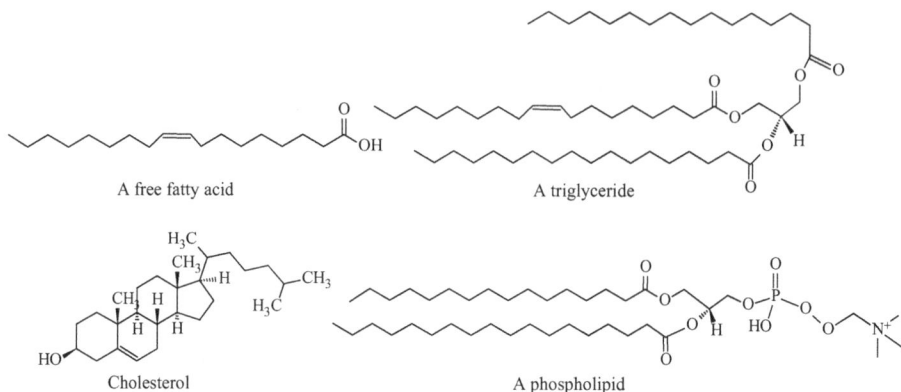

Figure 1.16: Structures of some common lipids.

1.11 Classification of Lipids

Lipids can be classified into different categories on the basis of structural composition and biological functions they perform in the living organisms. Based on structural composition, lipids can be categorized into three major groups:
1. Simple lipids
2. Complex Lipids
3. Derived lipids

1. Simple lipids: These are esters of fatty acids with different alcohols. They are further classified into fats, oils, and waxes.
2. Complex lipids: These are also known as compound lipids, which usually contain some additional groups such as carbohydrates, proteins, and phosphates in addition to fatty acid and alcohol. Phospholipids, glycolipids, and sphingolipids fall under the category of complex lipids.
3. Derived lipids: These can be obtained from the simple and compound lipids by means of hydrolysis. Fatty acids, glycerol, steroids, alcohols in addition to glycerol and sterols, fatty aldehyde and ketone bodies, cholesterol, bile acids, lipid-soluble vitamins, and hormones are examples of derived lipids.
    The most important lipids such as oils, fats, and waxes are the natural products that are extensively distributed in nature. Fats and oils also form an essential part of human diet along with proteins and carbohydrates. A small amount of fat is required for the healthy functioning of the body. Oils and fats provide calories, and essential fat also helps in absorption of fat-soluble vitamins such as A, D, E, and K. Oils and fats are composed of triglycerides (triacylglycerols), which are made up of three

molecules of fatty acids and one molecule of glycerol (Figure 1.17). Waxes are mixtures of esters of long-chain fatty acids with long-chain monohydric alcohols (one hydroxyl group). For example, cetyl palmitate, also known as hexadecyl hexadecanoate, is the ester derived from palmitic acid (hexadecenoic acid, $CH_3(CH_2)_{14}COOH$) and cetyl alcohol (hexadecanol, $CH_3(CH_2)_{15}OH$). Some common waxes are beeswax, carnauba wax, candelilla wax, and so on.

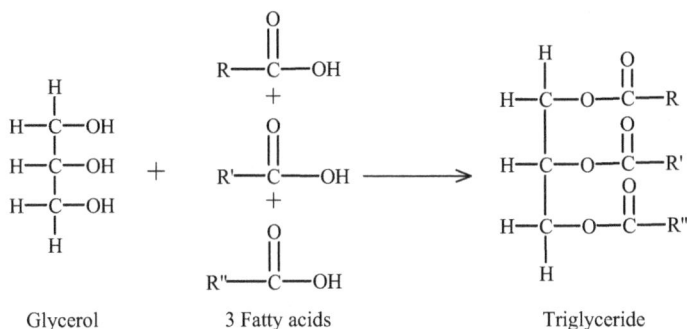

Figure 1.17: Formation of a triglyceride.

1.11.1 Fatty Acids Are Aliphatic Carboxylic Acids

Most lipids exhibit their hydrophobic characteristics due to their fatty acid component. Fatty acids consist of hydrocarbon chains of different lengths and degree of unsaturation with a terminal carboxyl group (carboxylic acid). These molecules usually contain an even number of carbons atoms in their hydrocarbon chains; mainly between 14 and 24, and the most common and widely distributed fatty acids in biological systems are 16- and 18-carbon fatty acids. Fatty acids are mainly present in the esterified form in oils and fats; however, in plasma, they also occur in unesterified form. Fatty acids that contain one or more double bonds in their hydrocarbon tails are described as unsaturated fatty acids, while those that lack any double bonds are known as saturated fatty acids.

There are different ways of naming a fatty acid; for example, they are either called by their systematic names according to the IUPAC (International Union of Pure and Applied Chemistry) nomenclature, by common names (trivial names), or by delta-x (Δ^x) nomenclature. The systematic name for a fatty acid is based on the numbering of carbon atoms, starting with the carboxylic (acidic) carbon, whereas common names specify their origin. For example, the systematic and common name of C_{16} saturated fatty acid is hexadecanoic acid and palmitic acid, respectively. A fatty acid containing 16-carbon atoms (C_{16}) with one double bond is called hexadecenoic acid; with two double bonds is called hexadecadienoic acid; and with three double bonds is called hexadecatrienoic acid. The notation 16:0 denotes a C_{16} fatty acid with no double bonds, whereas 16:2 indicates the presence of two double bonds. In delta-x (Δ^x) no-menclature, each double bond is indicated by Δ^x, where x indicates the position of the double bond from carboxylic end of the fatty acid. Each double bond is preceded by a cis- or trans- prefix, indicating the configuration of the molecule around the bond, and in most unsaturated fatty acids, the configuration is usually cis. For example, cis-Δ^9 depicts the presence of a cis-double bond between carbon atoms 9 and 10, and trans-Δ^{12} means there exists a trans double bond between carbon atoms 12 and 13. According to the IUPAC nomenclature, fatty acid carbon atoms are numbered from the carboxyl terminus, and carbon atoms 2 and 3 are named as α and β, respectively. The methyl carbon atom (-CH_3) present at the distal end of the fatty acid chain is re-ferred to as the ω-carbon (omega-carbon) atom. The position of a double bond can also be depicted by counting from the distal end, considering the ω-carbon atom as number 1. Fatty acids can be categorized in the omega groups such as omega-3, omega-6, and omega-9 fatty acids according to the location of their first double bond. The structures of the ionized form of stearic acid ($C_{18}H_{36}O_2$ (18:0)), oleic acid ($C_{18}H_{34}O_2$ (18:1)), and an ω-3 fatty acid are shown in Figure 1.18.

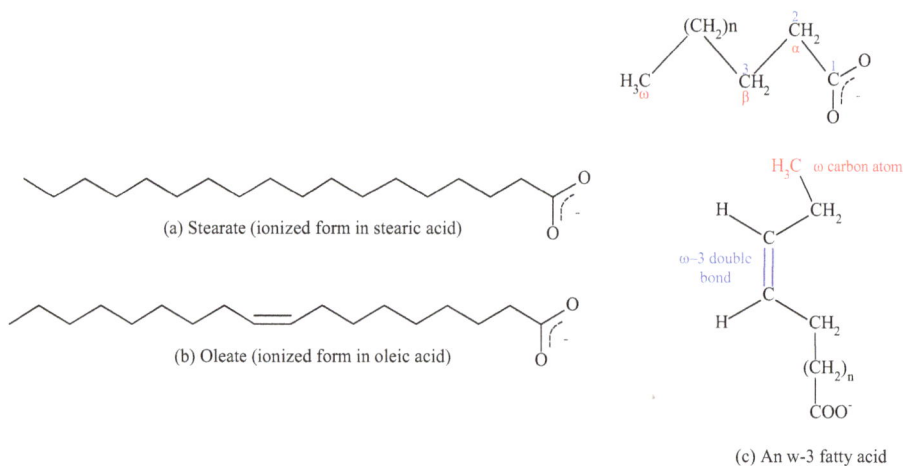

(a) Stearate (ionized form in stearic acid)

(b) Oleate (ionized form in oleic acid)

(c) An w-3 fatty acid

Figure 1.18: Structure of fatty acids. (a) Stearate is an 18-carbon, saturated fatty acid, (b) oleate is an 18-carbon fatty acid with a single cis double bond (between carbon 9 and 10), and (c) an ω-3 fatty acid.

A fatty acid is termed as saturated if there are no double bonds between carbon atoms of the chain; examples are palmitic acid (16 carbons) and stearic acid (18 car-bons). Unsaturated fatty acids may contain one (mono) or more (poly) double bonds between the carbon atoms of the molecular chain. They may further be divided into two groups on the basis of the number of double bonds – monosaturated fatty acids – having one double bond, and polyunsaturated fatty acids – having more than one double bond. The 18-carbon unsaturated fatty acids containing one, two, and three double bonds are called oleic acid, linoleic acid, and linolenic acid, respectively. Some of the saturated and unsaturated fatty acids are given in Table 1.1.

Table 1.1: Some naturally occurring fatty acids.

Number of carbon atoms	Number of double bonds	Common name	Formula	Systematic name
10	0	Caprate	$CH_3(CH_2)_8COO^-$	n-Decanoate
12	0	Laurate	$CH_3(CH_2)_{10}COO^-$	n-Dodecanoate
14	0	Myristate	$CH_3(CH_2)_{12}COO^-$	n-Tetradecanoate
16	0	Palmitate	$CH_3(CH_2)_{14}COO^-$	n-Hexadecanoate
18	0	Stearate	$CH_3(CH_2)_{16}COO^-$	n-Octadecanoate
20	0	Arachidate	$CH_3(CH_2)_{18}COO^-$	n-Eicosanoate
22	0	Behenate	$CH_3(CH_2)_{20}COO^-$	n-Docosanoate
24	0	Lignocerate	$CH_3(CH_2)_{22}COO^-$	n-Tetracosanote
16	1	Palmitoleate	$CH_3(CH_2)_5CH = CH(CH_2)_7COO^-$	cis-Δ^9-Hexadecenoate
18	1	Oleate	$CH_3(CH_2)_7CH = CH(CH_2)_7COO^-$	cis-Δ^9-Octadecenoate
18	2	Linoleate	$CH_3(CH_2)_4(CH = CHCH_2)_2(CH)_6COO^-$	cis, cis-Δ^9, Δ^{12}-Octadecadienoate
18	3	Linolenate	$CH_3CH_2(CH = CHCH_2)_3(CH_2)_6COO^-$	all cis-$\Delta^9, \Delta^{12}, \Delta^{15}$-Octadecatrienoate
20	4	Arachidonate	$CH_3(CH_2)_4(CH = CHCH_2)_4(CH_2)_2COO^-$	all cis-$\Delta^5, \Delta^8, \Delta^{11}, \Delta^{14}$-Eicosatetraenoate

In naturally occurring unsaturated long-chain fatty acids, the configuration at the double bond is usually cis rather than trans. The difference between cis- and trans-configuration is quite important in determining the overall shape of the fatty acid. A trans-fatty acid has a shape similar to that of a saturated fatty acid, whereas a cis-double bond puts a kink in the hydrocarbon chain. When number of cis-double bonds increase in a fatty acid, it results in numerous possible spatial configurations; for example, arachidonic acid with four double cis-bonds has kink and attains U shape. This might have a pivotal role in molecular assemblage in membranes and on the positions acquired by these fatty acids in complex biopolymers such as phospholipids, whereas the presence of trans-bond at such positions hamper these spatial arrangements.

The saturated fatty acid carbon chains can adopt different conformations but they inclined to remain fully extended as this curtails the steric repulsion among adjacent methylene groups. The uniform fatty acid chains in saturated fatty acids allow the molecules to pack tightly and efficiently into crystals; as a result, they exhibit higher melting points, and the melting points increase as the molecular weight increases. The cis-configuration of unsaturated fatty acids generates a rigid bend in the molecular chain that disrupts the packing and decreases the dispersion force attraction between molecules, resulting in a lower melting point in comparison to saturated fatty acids.

The properties of various fatty acids depend on the chain length and the degree of unsaturation. Fats are triglycerides that are composed of fully saturated fatty acids and are solids at room temperature. Triglycerides consisting of unsaturated fatty acids likely to exist as liquids at room temperature are oils. In fats, the saturated fatty acid chain can pack close to each other, causing them to be solids at room temperature, while in case of oils, the unsaturated fatty acid chains cannot pack tightly together; hence, they are liquids and have usually low melting points. Fats are mostly obtained from animals, whereas triglycerides originated from plants are oils. Therefore, we commonly speak of vegetable oil and animal fat. Each type of oil and fat has a different combination of fatty acids, and the nature of the fatty acid determines the consistency of the oil and fat. Fatty acid composition of different fats and oils is given in Table 1.2.

Depending on the nature of fatty acid, triglycerides can be classified as simple or mixed triglycerides. Simple triglycerides are formed when all three hydroxyl (-OH) groups on the glycerol are esterified with the same fatty acid, whereas mixed triglycerides are obtained from esterification of glycerol with two or three different fatty acids. The chemical and physical properties of oils and fats depend on the composition of fatty acids. Oils and fats are insoluble in water but soluble in organic solvents like hexane, benzene, carbon tetrachloride, petroleum ether, and carbon disulfide. When pure, they are colorless, odorless, and tasteless, and have low densities than water. Oils and fats exhibit lubricating properties, and are greasy to touch.

Essential fatty acids (EFAs) are those fatty acids which are required for the proper growth of the body but cannot be synthesized by human body, so these must be taken in our diet. Linoleic acid, an omega-6 fatty acid, and alpha-linolenic acid, an omega-3 fatty acid, are the two EFAs known for humans. On the other hand, fatty acids which can be synthesized by our body are known as non-EFAs, and thus, they need not be included in our diet. It is important to note that non-EFAs do not mean that these fatty acids are not important, the categorization is based solely on the ability of human body to synthesize that particular fatty acid. EFAs are required for the proper development and functioning of the brain and nervous system as well as for regulation of blood pressure. Flaxseed oil, fish, walnuts, hemp, and leafy vegetables are some of the excellent sources of omega-3 and omega-6 EFAs. EFA deficiency is extremely rare as these fatty acids are easily accessible.

Table 1.2: Composition of fatty acid in some common edible fats and oils.

Oil of fat	Unsaturated/ saturated ratio	Saturated					Monounsaturated (MUFA)	Polyunsaturated (PUFA)	
		Capric acid C10:0	Lauric acid C12:0	Myristic acid C14:0	Palmitic acid C16:0	Stearic acid C18:0	Oleic acid C18:1	Linoleic acid C18:2	Alpha Linolenic acid C18:3
Beef Tallow	0.9	–	–	3	24	19	43	3	1
Butterfat (cow)	0.5	3	3	11	27	12	29	2	1
Butterfat (human)	1.0	2	5	8	25	8	35	9	1
Canola oil	15.7	–	–	–	4	2	62	22	10
Cocoa butter	0.6	–	–	–	25	38	32	3	–
Cod liver oil	2.9	–	–	8	17	–	22	5	–
Coconut oil	0.1	6	47	18	9	3	6	2	–
Corn oil (maize oil)	6.7	–	–	–	11	2	28	58	1
Cottonseed oil	2.8	–	–	1	22	3	19	54	1
Flaxseed oil	9.0	–	–	–	3	7	21	16	53
Grapeseed oil	7.3	–	–	–	8	4	15	73	–
Lard (pork fat)	1.2	–	–	2	26	14	44	10	–
Olive oil	4.6	–	–	–	13	3	71	10	1

(continued)

Table 1.2 (continued)

Oil of fat	Unsaturated/ saturated ratio	Saturated					Monounsaturated (MUFA)	Polyunsaturated (PUFA)	
		Capric acid C10:0	Lauric acid C12:0	Myristic acid C14:0	Palmitic acid C16:0	Stearic acid C18:0	Oleic acid C18:1	Linoleic acid C18:2	Alpha Linolenic acid C18:3
Palm oil	1.0	–	–	1	45	4	40	10	–
Palm kernel oil	0.2	4	48	16	8	3	15	2	–
Peanut oil	4.0	–	–	–	11	2	48	32	–
Safflower oil	10.1	–	–	–	7	2	13	78	–
Sesame oil	6.6	–	–	–	9	4	41	45	–
Soybean oil	5.7	–	–	–	11	4	24	54	7
Sunflower oil	7.3	–	–	–	7	5	19	68	1
Walnut oil	5.3	–	–	–	11	5	28	51	5

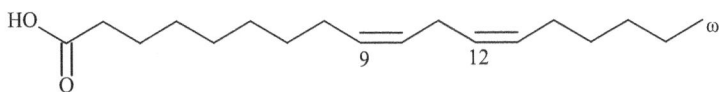

Linoleic acid (9, 12 Octadecadienoic acid) (omega-6 fatty acid)

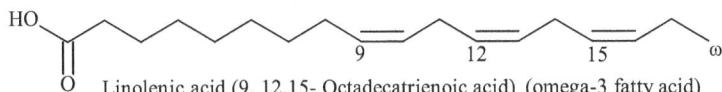

Linolenic acid (9, 12,15- Octadecatrienoic acid) (omega-3 fatty acid)

1.12 Oils and Fats: Chemical and Physical Properties

The important physicochemical properties of fats and oils are melting point, polymorphism, fatty acid composition, and solid fat content. We know that triglycerides, the main constituents of body fat in humans, and other vertebrates are stored in adipose tissues. The reactions of oils and fats are basically reactions of triglycerides. They undergo hydrolysis yielding glycerol and three molecules of fatty acids. In biological systems, triglycerides are hydrolyzed by lipases, enzymes present in the digestive tract of human beings and animals that degrade triglycerides to free fatty acids (FFAs) and glycerol. When hydrolysis of triglycerides is carried out under alkaline medium, it leads to the formation of sodium and potassium salts of fatty acids, called soaps, and this process of soap formation is called saponification.

The prolonged exposure of oils and fats to moisture, air, light, and higher temperature results in deterioration of their quality, and this process is known as rancidity. The hydrolysis and oxidation reactions are responsible for the unpleasant taste and odor associated with oils and fats. The process of hydrolysis generates FFAs and can be determined through titration with the standard solution of a base. Higher acid value is an indicator of poor quality of the oil/fat. Some of the important parameters such as hydrogenation, saponification value, iodine number, and acid value of fats and oils are discussed further:

i. Hydrogenation of fats and oils: Triglycerides consisting of polyunsaturated fatty acids (vegetable oils) are less stable and turn rancid by autoxidation more quickly in comparison to animal fats. Therefore, oils are converted into solid fats by adding hydrogen (in a process called hydrogenation) which raises the melting point of the triglycerides, change the consistency and texture of the product, prevents oxidation of oils, and increase their shelf life. Hydrogenation is a chemical process in which hydrogen is added to natural unsaturated fats, such as vegetable oil, to turn it into a solid fat at room temperature in the presence of a metal catalyst, usually nickel. This process is similar to the catalytic hydrogenation of alkenes, decreases the number of double bonds, and hence, retards and eliminates the potential of an oil to get rancid. An example of hydrogenation is the production of margarine and shortening. Vegetable oil is too soft for margarine or shortening, whereas saturated fat is too hard. Margarine requires something that is neither very hard nor very soft, and that desired consistency can be obtained by partial hydrogenation of vegetable oils. Glyceryl trioleate upon complete hydrogenation yields glyceryl tristearate:

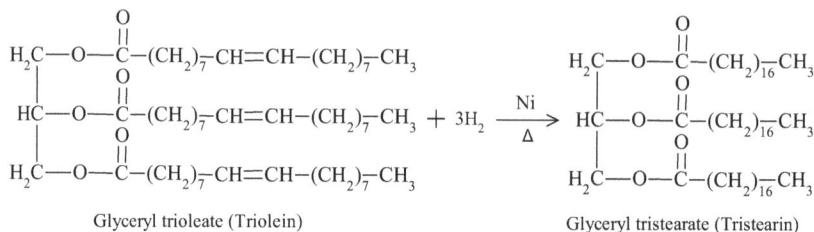

$$
\begin{array}{l}
\text{H}_2\text{C}-\text{O}-\overset{\overset{\displaystyle O}{\|}}{\text{C}}-(\text{CH}_2)_7-\text{CH}=\text{CH}-(\text{CH}_2)_7-\text{CH}_3 \\[4pt]
\text{HC}-\text{O}-\overset{\overset{\displaystyle O}{\|}}{\text{C}}-(\text{CH}_2)_7-\text{CH}=\text{CH}-(\text{CH}_2)_7-\text{CH}_3 \;+\; 3\text{H}_2 \\[4pt]
\text{H}_2\text{C}-\text{O}-\overset{\overset{\displaystyle O}{\|}}{\text{C}}-(\text{CH}_2)_7-\text{CH}=\text{CH}-(\text{CH}_2)_7-\text{CH}_3
\end{array}
\quad\xrightarrow[\Delta]{\text{Ni}}\quad
\begin{array}{l}
\text{H}_2\text{C}-\text{O}-\overset{\overset{\displaystyle O}{\|}}{\text{C}}-(\text{CH}_2)_{16}\text{CH}_3 \\[4pt]
\text{HC}-\text{O}-\overset{\overset{\displaystyle O}{\|}}{\text{C}}-(\text{CH}_2)_{16}\text{CH}_3 \\[4pt]
\text{H}_2\text{C}-\text{O}-\overset{\overset{\displaystyle O}{\|}}{\text{C}}-(\text{CH}_2)_{16}\text{CH}_3
\end{array}
$$

Glyceryl trioleate (Triolein)   Glyceryl tristearate (Tristearin)

Hydrogenation is of great commercial importance, and hydrogenated vegetable oils are used in the food industry to improve the taste and texture of products. Commercial cooking fats are prepared by the partial hydrogenation of vegetable oils, which results in "partially hydrogenated fats" as seen in many food items. When complete hydrogenation of polyunsaturated fat is carried out, all the double bonds are converted into the saturated ones, but complete hydrogenation of the oil is usually avoided because a fully saturated triglyceride is quite hard and brittle. So, hydrogenation of vegetable oil is carried out until a semisolid consistency is attained. Now the question arises, do hydrogenated vegetable oils been linked to several health effects? A main health concern during the hydrogenation of liquid vegetable oils is the production of trans fat. Trans fats are present in certain foods, and they arise as a byproduct during the saturation of fatty acids in the process of hydrogenation or "hardening" of vegetable oils in the production of margarine. Unlike other dietary fats, trans fat is considered the worst type of fat to eat; it raises "bad" cholesterol and also lowers "good" cholesterol. A diet laden with trans fats increases the risk of heart diseases and stroke. According to a regulation issued by the Food and Drug Administration (FDA) in 2003, manufacturers are required to list trans fat in the nutrition label of conventional foods and dietary supplements. With such information in nutrition labeling, consumers can make healthier choice and limit their consumption of trans fat which helps them in maintaining healthy dietary practices.

ii. Saponification value: The saponification value is one of the main important characteristic values of fat. The ester bonds present in triglycerides (oils and fats) can be hydrolyzed under acidic or basic conditions or with the help of enzyme lipases under biological conditions. The process of hydrolysis of triglycerides under alkaline medium (usually potassium hydroxide, KOH) is called saponification, and this results in the formation of glycerol and a mixture of salts (potassium) of fatty acids. These salts of long-chain fatty acid are called soaps, and most soaps are prepared in this way. Saponification value is defined as the number of milligrams of KOH required to saponify 1 g of fat or oil. Saponification value of a fat or oil can be determined by refluxing it with excess of alcoholic KOH solution until hydrolysis is complete. The sample is saponified, that is, the esters are split into fatty acids and glycerol. The excess of KOH can then be titrated with standardized hydrochloric acid to estimate how much alkali was required for saponification of the sample.

Since three ester bonds are present in any triglyceride molecule to hydrolyze, three equivalents of KOH are needed to saponify one molecule of any fat or oil. Thus, saponification number can be calculated by taking the example of glyceryl tripalmitate:

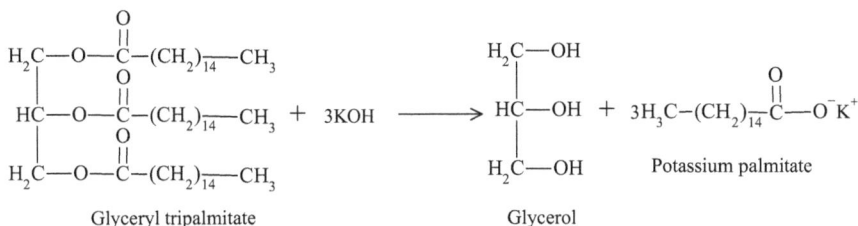

$$H_2C-O-\overset{\overset{O}{\|}}{C}-(CH_2)_{\overline{14}}-CH_3$$
$$HC-O-\overset{\overset{O}{\|}}{C}-(CH_2)_{\overline{14}}-CH_3 \;+\; 3KOH \longrightarrow$$
$$H_2C-O-\overset{\overset{O}{\|}}{C}-(CH_2)_{\overline{14}}-CH_3$$

Glyceryl tripalmitate

$$H_2C-OH$$
$$HC-OH \;+\; 3H_3C-(CH_2)_{\overline{14}}\overset{\overset{O}{\|}}{C}-O^-K^+$$
$$H_2C-OH$$

Glycerol

Potassium palmitate

Molecular weight of glyceryl tripalmitate (tripalmitin) = 807.32 g/mol
Three equivalents of KOH are required to completely saponify
That is, 807.32 g of fat = 168.32 g of KOH
$$= 168,320 \text{ mg of KOH}$$
Therefore,

$$1\,g \text{ of fat will require } \frac{168,320}{807.32} \text{ mg of KOH}$$

∴ Saponification number of glyceryl tripalmitate is 208.5.

Similarly, we can determine the saponification number of any fat or oil.

iii. Iodine number: Oils and fats contain different types of saturated and mono/poly-unsaturated fatty acids. Iodine number is an important characteristic of oils that indicates the amount of unsaturated fatty acids. Iodine number or iodine value is defined as the number of grams of iodine consumed by 100 g of fat or oil sample for saturation. For a saturated triglyceride, the iodine value is zero. Thus, for a fat iodine value is low, whereas it is high for an oil. Higher iodine value indicates the higher degree of unsaturation in an oil sample, and hence, more is the possibility of the sample to get rancid.

Iodine value can be determined by dissolving a known quantity of fat/oil in a suitable solvent, such as chloroform, and allowed to react with an excess of iodine monochloride. As iodine does not react directly with an oil sample, Wijs reagent (iodine monochloride in acetic acid) is used. The sample mixture is then treated with potassium iodide to liberate free iodine, which can be estimated by titrating with standardized sodium thiosulfate solution using starch as an indicator. Blank titration is also performed at the same time to determine the amount of iodine consumed by the sample under analysis. Here, we have taken the example of glyceryl trioleate (triolein) to calculate its iodine number. Triolein reacts with three molecules of iodine to give the addition product, as shown. The molecular weight of triolein is 885.432 g/mol:

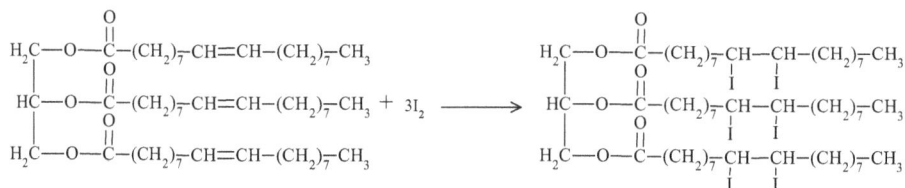

This reaction shows that 761.42 g (126.90 × 6) of iodine is added to 885.432 g of triolein.

The number of grams of iodine that is added to 100 g of triolein will be $\dfrac{761.42 \times 100}{885.432}$

∴ Iodine number of triolein is 86.

iv. Acid value: Acid value is an important parameter that determines the number of FFA in an oil or fat, which have been liberated by hydrolysis from the triglycerides due to the action of moisture or temperature or enzyme. Acid value is defined as the number of milligrams of KOH required to neutralize the FFAs present in 1 g of fat or oil sample. It can be estimated by dissolving a known amount of oil or fat in a neutral solvent (equal volume of alcohol and diethyl ether) and titrating it against standard KOH using phenolphthalein as an indicator. Acid value indicates the degree of rancidity, and often an important parameter in evaluating the deterioration of oil products. Cooking oils and petroleum products often get degraded over time as their chemical constituents dissociate and release hydrogen ions, resulting in higher acid values of rancified oils. Degradation of oils and fats is a concern for the food industry because increased acidity affects taste and quality, which may pose health risks. Rancidification cannot be eliminated completely; however, it can be reduced by storing the fats and oils at low temperature and away from light.

1.13 Rancidity

The deterioration of oils and fats associated with their storage has been a major concern since ancient times. The edible oils, fats, and their food products get degraded on storage which is indicated by the development of off-flavor, off-odor, sometimes, accompanied by a change in taste and appearance. This change occurs when oils, fats, and their foodstuffs come in contact with atmospheric oxygen/moisture. The action of enzyme and microorganisms can also alter the structure of oils and fats. This natural process of decomposition of oils and fats by either hydrolysis or oxidation or both results in the development of off-flavor and off-odor, and change in color and taste is referred to as rancidity. The majority of solid foodstuff and vegetable oils consist of unsaturated fatty acid, and hence they turn rancid more quickly than do animal fats. The carbon–carbon double bonds present in unsaturated fatty acids can easily undergo oxidation, resulting in the shortening of hydrocarbon chains along with the formation of oxygenated products that taste and smell different. As unsaturated fats are more prone to oxidation than saturated fats, more the number of polyunsaturated fatty acid present in a sample, higher will be their tendency to go rancid. This is due to the presence of more number of unstable double bonds, which allow more oxygen to react at those points.

There are two main types or causes of rancidity that are responsible for the degradation of stored edible oils: oxidative and hydrolytic.

i. Oxidative rancidity: Oxidative rancidity, also known as autoxidation, occurs when aerial oxygen is absorbed. Such a rancidity develops, sooner or later, when substances with substantial of glycerides of fatty acids are exposed to atmospheric oxygen. During this process, oxygen molecules interact with the structure of the oils, and degrade into compounds such as fatty acids, aldehydes, and ketones. These compounds with shorter carbon chains change the color, odor, and taste, and makes them unfit for consumption. Oxidative rancidity not only causes undesirable flavors and odors but also leads to the formation of substances such as peroxidized fatty acids, polymeric material, and oxidized sterols that exert harmful nutritional and physiological effects. These substances destroy fat-soluble vitamins and thought to be involved in atherosclerosis.

ii. Hydrolytic rancidity: Hydrolytic rancidity is caused when water hydrolyzes the bonds in triglycerides, resulting in the generation of an FFA, diglycerides (diacylglycerols, DAGs), monoglycerides (monoacylglycerols, MAGs), and glycerol. All these degradation products are responsible for causing undesirable change in the quality of an oil/fat so that the odor and flavor of oil/fat become unpleasant. The hydrolysis reaction that causes rancidity does not occur as easily as the oxidative reaction. In most cases, hydrolytic rancidity is catalyzed by lipases, which is present in both animal and plant tissues. The presence of this enzyme liberates lower fatty acids by hydrolyzing the ester bond which is responsible for a strong smell and rancidity. For example, butyric acid in butter contributes to that off flavor of butter. Some microorganisms also contain enzymes that may generate oxidation products from fatty acid esters, which become responsible for hydrolytic rancidity, for example, development of rancid flavor in milk.

Action of heat, enzyme, and moisture

Partial or complete hydrolysis of ester bonds ⟶ Liberation of FFA, MAGs, DAG, glycerol

$$
\begin{array}{c}
\mathrm{H_2C-O-\overset{\displaystyle O}{\overset{\|}{C}}-R} \\[4pt]
\mathrm{HC-O-\overset{\displaystyle O}{\overset{\|}{C}}-R^1} \\[4pt]
\mathrm{H_2C-O-\overset{\displaystyle O}{\overset{\|}{C}}-R^2}
\end{array}
\longrightarrow
\begin{array}{c}
\mathrm{H_2C-O-\overset{\displaystyle O}{\overset{\|}{C}}-R} \\[4pt]
\mathrm{HC-O-\overset{\displaystyle O}{\overset{\|}{C}}-R^1} \\[4pt]
\mathrm{H_2C-OH}
\end{array}
\longrightarrow
\begin{array}{c}
\mathrm{H_2C-OH} \\[4pt]
\mathrm{HC-O-\overset{\displaystyle O}{\overset{\|}{C}}-R^1} \\[4pt]
\mathrm{H_2C-OH}
\end{array}
\longrightarrow
\begin{array}{c}
\text{Free Fatty Acids} \\
+ \\
\text{Glycerol}
\end{array}
$$

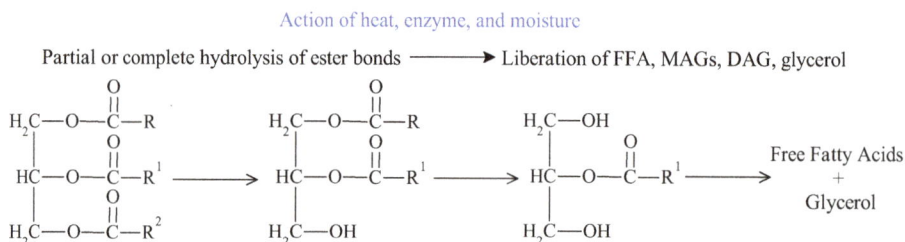

1.13.1 Prevention of Rancidity

It is very important to preserve the food properly to avoid rancidity in the fat/oil-containing food items and to maintain their desired quality. Antioxidants are the most effective way to keep food from getting rancid and to prevent the autoxidation action in food items containing fats and oils. Antioxidants can either be natural or synthetic; for example, flavonoids, polyphenols, vitamin C, and tocopherols are all natural antioxidants, whereas butylated hydroxytoluene, butylated hydroxy anisole, propyl 3,4,5-trihydroxybenzoate (also known as propyl gallate) are some of the examples of synthetic antioxidants. Sequestering agents such as EDTA (ethylenediaminetetraacetic acid) and citric acid can also prevent oxidation, and hence, rancidity.

Another way to prevent the food from becoming rancid is to keep it in a cold and dark place, that is, away from light and air. The food items susceptible to rancidity can be stored in airtight containers. Do not add new oil to container that already has oil in them as old oil will cause a reaction, and making the new oil to get rancid faster.

1.14 Reversion

Many oils and fats develop a change in flavor before turning rancid, and this change in flavor, which is quite different from the rancid flavor, is termed as reversion. Reversion is defined as a change in flavor of edible fats/oils that are characterized by the development of an unpleasant flavor in prior to the onset of actual rancidity. Various factors such as the presence of small amounts of metal ions like copper and iron, exposure of fat to ultraviolet or visible light, and heat affect the flavor reversion. In case of rancidity, the change in flavor is same for all fats/oils, whereas in reversion, it can be grassy, buttery, fishy, painty, or beany depending on the oil which reverts. It occurs principally in fish oils, linseed oils, and soybean oils, which contain fatty acids with three or more double bonds, whereas rancidity can be observed in all oils.

1.15 Complex lipids

1.15.1 Phospholipids

Phospholipids are abundant in all biological membranes and are composed of four components: one or more fatty acids, glycerol or sphingosine (a more complex alcohol) to which fatty acids are attached, a phosphate, and an alcohol attached to the phosphate. Phospholipids which are derived from glycerol are called phosphoglycerides. In phosphoglycerides, the C-1 and C-2 hydroxyl groups of glycerol are esterified to the carboxyl group of fatty acids, and the hydroxyl group at C-3 is esterified to phosphoric acid. The resulting compound is known as phosphatidic acid (Figure 1.19). Their molecular structure is polar, having one hydrophilic head and two hydrophobic tails. They constitute the major component of lipid bilayer of cell membranes. The structure is called a "lipid bilayer" because it is made up of two layers of phospholipids. The polar heads of phospholipids are on the outside of the cell, whereas their hydrophobic tails (fatty acids) form the interior of the bilayer. A second layer of phospholipids also forms with their heads facing the inside of the cell and tails facing away (Figure 1.20). Lipid bilayer is unique in the fact that polar head (hydrophilic head) interacts with water, and the fluidity of membrane is maintained and controlled by the fatty acid components of phospholipids. The presence of unsaturated fatty acids enhances membrane fluidity because they are not so closely packed, whereas in long-chain saturated fatty acids, the hydrocarbon chains are closely packed together which decreases fluidity.

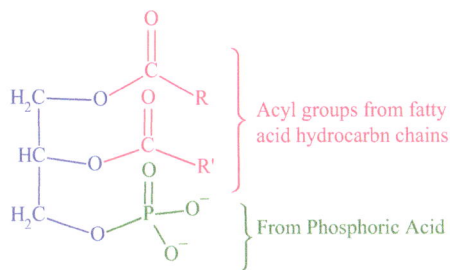

Figure 1.19: Structure of phosphatidate.

Some major phosphoglycerides are obtained when the phosphate group of phosphatidate is esterified with the hydroxyl group of some alcohols such as serine, ethanolamine, choline, and inositol resulting in the formation of phosphatidylserine, phosphatidylethanolamine, phosphatidylcholine, and phosphatidylinositol, respectively, as shown in Figure 1.21.

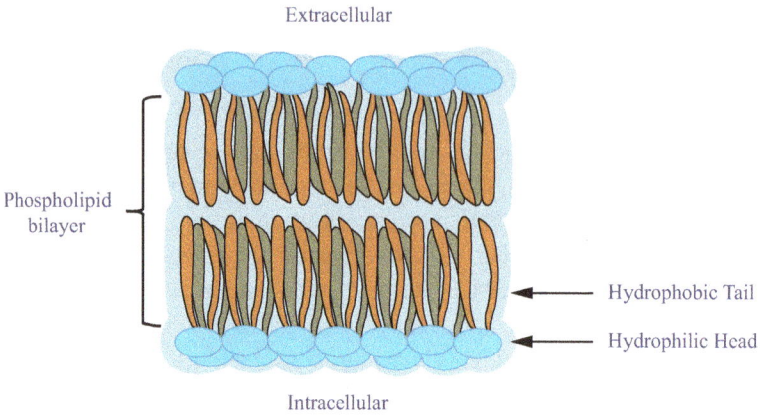

Figure 1.20: Structure of lipid bilayer in cell membrane.

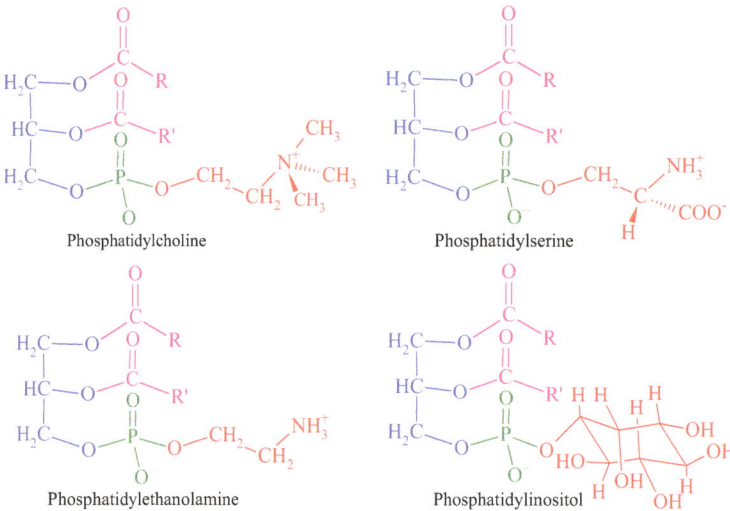

Figure 1.21: Structure of some of the common phosphoglycerides found in membranes.

1.15.2 Sphingolipids

Sphingolipids encompass a vast range of complex lipids in which sphingosine, an amino alcohol consisting of a long unsaturated hydrocarbon chain, is linked to a fatty acid via an amide bond. They are found in both animals and plants, and in some prokaryotic organisms, fungi, and viruses. Sphingolipids play an important role in signal transduction and cell recognition. The simplest compound of this class is ceramides, which consists of one fatty acid attached to the amino group of sphingosine via amide linkage. Sphingomyelin is the most abundant complex sphingolipid found in animal cell membranes. In sphingomyelin, the primary hydroxyl group of the sphingosine is esterified to phosphorylcholine, and an amino group is attached to a fatty acid via an amide bond. The structures of sphingosine and sphingomyelin are shown in Figure 1.22.

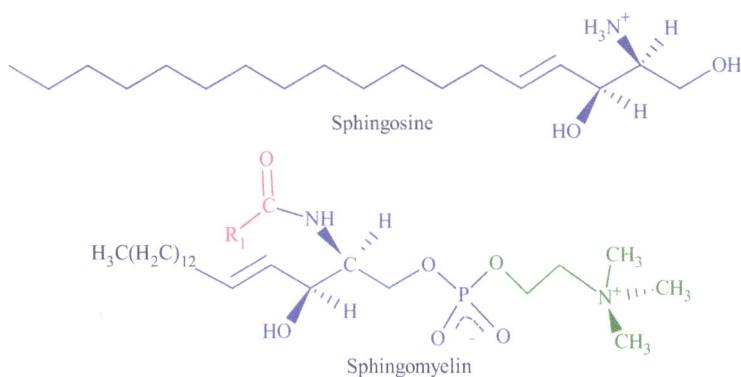

Figure 1.22: Structures of sphingosine and sphingomyelin.

1.15.3 Glycolipids

Glycolipids, sugar-containing lipids, are the second major class of membrane lipids. Glycolipids are formed when a carbohydrate, monosaccharide or oligosaccharide, is attached to the alcohol group of a lipid by a glycosidic bond. It includes a wide variety of compounds such as glycosphingolipids (cerebrosides, gangliosides, globosides, etc.), glycoglycerolipids, and glycophospholipids. Cerebrosides, an important constituent of the brain and other tissues, consists of ceramide (sphingosine and a fatty acid) and a monosaccharide, usually glucose or galactose. The structure of glucocerebroside, a glycolipid consists of a glucose molecule, is shown in Figure 1.23, and it plays an important role in maintaining the stability of the cell membranes.

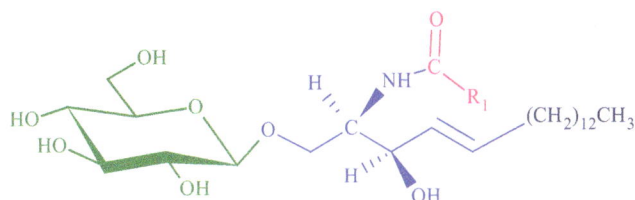

Figure 1.23: Structure of a glucocerebroside.

1.16 Triglycerides

A triglyceride molecule (triacylglycerol, TAG, or triacylglyceride) is an ester consisting of a glycerol and three fatty acids (Figure 1.24). Triglycerides are the main constituents of vegetable oils and animal fats in our diet. We get triglycerides from two sources – fats we eat in the diet and our body makes them from carbohydrates. When we consume extra calories, especially carbohydrates – liver increases the production of triglycerides. Triglycerides epitomize the main form of lipid storage and energy in humans. When we consume more energy or calories than needed, our body stores that in the fat cells in the form of triglycerides for later use, and when required, releases them as fatty acids, which fuel body movements and provide energy for important biological functions, and physical activity. The enzyme pancreatic lipase hydrolyzes the ester bond in the triglycerides, and thus releases the fatty acid. Once the triglycerides have been broken down, fatty acids, monoglycerides, and some diglycerides are then absorbed by the duodenum.

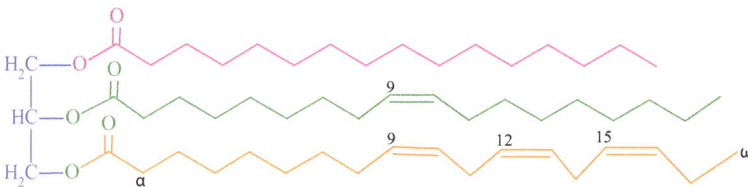

Figure 1.24: Structure of a triglyceride (a triacylglycerol).

Triglyceride is an important source of energy stored in fat tissues. It also works as an insulation material. It provides a thermal and mechanical protective layer that keeps humans and animals warm by being located underneath the skin, and this insulating layer gets reduced with aging. Although consuming lipids is a dietary necessity but taking these compounds in excess amount can have adverse effect on our health. The consumption of excess quantity of fat leads to the accumulation of triglycerides in fat cells which results in weight gain. Studies have shown that the high levels of triglycerides may result in pancreatitis, hardening of arteries, and increases the risk of stroke, coronary artery disease, and heart attack. The level of triglyceride can be measured by taking a blood sample, and results reveal whether your triglycerides fall into a healthy range, or not. For your good health, the triglyceride level should be less than 150 mg/dL (milligrams per deciliter); borderline high levels are from 151 to 199 mg/dL; high level ranges between 200 and 499 mg/dL; and more than 500 mg/dL falls in the very high range.

The factors that may contribute to high blood triglyceride levels can be genetic or due to excessive alcohol intake, liver or kidney diseases, some medications including hormones, diuretics, corticosteroids, thyroid problems, uncontrolled diabetes, and a high intake of sugar. One can lower the triglyceride number by adopting a healthy diet and staying active. The normal triglyceride level can be maintained by consuming less saturated fats, limiting foods high in sugar, eating foods high in omega-3 fatty acids, including fruits, vegetables, and whole grains in our diet, and making exercise as a part of our routine.

1.17 Cholesterol – A Lipid Based on a Steroid Nucleus

Cholesterol is an important sterol, biosynthesized by all animal cells, and helps our body to perform various important functions. It is the third major type of membrane lipid, and has a unique structure that is distinctly different from other phospholipids. Cholesterol, a steroid containing 27 carbon atoms, is made up of 4 linked hydrocarbon rings, a hydrocarbon tail at one end of the steroid, and a hydroxyl group at the other end. Cholesterol is oriented parallel to the fatty acid chains of the phospholipids in cell membranes, where the hydroxyl group interacts with the polar head of the nearby phospholipid.

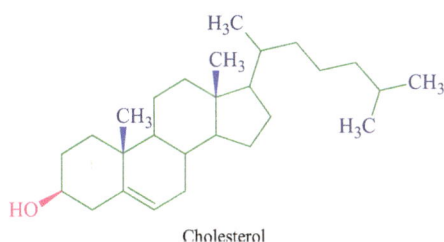

Cholesterol

Cholesterol and other fats are carried in the blood by proteins, and when these fats and proteins combine, they are called lipoproteins. The two main lipoproteins are:

- LDL (low-density lipoprotein) – This is often called *bad cholesterol*. This lipoprotein transports cholesterol from the liver to cell, but if too much is carried, more than the cells use, there can be a harmful accumulation of LDL in the walls of arteries, causing them to narrowing. With time, LDL can enhance the risk of arterial diseases if level goes too high.
- HDL (high-density lipoprotein) – This is also termed as *good cholesterol* because HDL carries excess cholesterol away from the cells to the liver where it breaks down and is removed from the body as waste.

We get some of the cholesterol from our diet but mainly it is made in our liver. It is an essential structural component of animal cell membranes. Cholesterol also functions as a precursor for the synthesis of steroid hormones such as cortisol, testosterone, estradiol, fat-soluble vitamin D, and bile. It also aids in the metabolism of fat-soluble vitamins (vitamins A, D, E, and K). It helps in regulating membrane permeability, functions as transporters and signaling molecules, insulates nerve fibers, and is also crucial for several biochemical pathways. But high level of cholesterol can increase the risk of heart diseases. Consumption of foods that are high in cholesterol, saturated fat, and trans fat (such as whole milk, butter, cheese, egg yolks, and meat) may increase the level of cholesterol. Some other factors that play a role in causing high cholesterol are sedentary lifestyle, smoking, tobacco and alcohol use, inactivity, family history, some medicines, and medical conditions. Stress is also one of the factors that triggers hormonal changes that results in producing high cholesterol. Unfortunately, a high cholesterol level does not cause any symptoms for most people, and a person would not feel any symptoms until the enhanced cholesterol causes other problems in the body. A blood test is the only way to detect it.

Increased cholesterol level in the blood can increase the risk of many medical problems such as atherosclerosis (narrowing of the arteries), peripheral artery/vascular disease, coronary heart diseases, angina, stroke and ministroke, and heart attack. One can maintain the cholesterol number on the lower side by adopting heart-healthy lifestyle, eating a diet rich in fruits, vegetables, and whole grains, avoiding saturated fats, quitting alcohol and smoking, and staying active.

References

[1]   Berg JM, Tymoczko JL, Gatto GJ Jr. Lubert Stryer. Biochemistry. Freeman Macmillan. 8th Edition. ISBN-13. 2006:978–1.
[2]   Bhutani SP. Chemistry of Biomolecules. CRC Press Boca Raton, Florida; 2019 Sep 25.
[3]   Cox RA, García-Palmieri MR. Cholesterol, Triglycerides, and Associated Lipoproteins. In: Walker HK, Hall WD, Hurst JW, editors. Clinical Methods: The History, Physical, and Laboratory Examinations. 3rd ed. Boston: Butterworths; 1990. Chapter 31. PMID: 21250192 Page no 153–160. Available from: https://www.ncbi.nlm.nih.gov/books/NBK351/.
[4]   Finar IL. Organic Chemistry (Volume 2), Dorling Kindersley (India) Pvt. Ltd.
[5]   Hames D, Hooper N. Instant Notes Biochemistry. Taylor & Francis; 2006 Sep 27.
[6]   https://www.huhs.edu/literature/EssentialFattyAcids.pdf
[7]   Marriott BP, Birt DF, Stalling VA, Yates AA, editors. Present Knowledge in Nutrition: Basic Nutrition and Metabolism. Academic Press; 2020 Jul 21. Peter J.H. Jones, Alice H. Lichtenstein, Chapter 4 - Lipids, Editor(s): Bernadette P. Marriott, Diane F. Birt, Virginia A. Stallings, Allison A. Yates, Present Knowledge in Nutrition (Eleventh Edition), Academic Press, 2020, Pages 51–69, ISBN 9780323661621, https://doi.org/10.1016/B978-0-323-66162-1.00004-4.
[8]   Mehta B, Mehta M. Organic Chemistry. PHI Learning Pvt. Ltd.; 2015 Aug 31.
[9]   Morrison RT. Organic Chemistry. Pearson Education India; 1972.
[10] Pelley JW, Goljan EF. Rapid Review Biochemistry E-Book.
[11] Watson H. Biological membranes. Essays in Biochemistry. 2015 Nov 15; 59:43–69.

Chapter 2
Amino Acids, Peptides, and Proteins

The term "Protein" was first coined by J. J. Berzelius in 1838, and derived from the Greek word "proteios" meaning holding first place. Proteins are the highly complex biomolecules in living systems, and centrally involved in essentially all biological processes. These macromolecular structures are made up of a chain of hundreds or thousands of α-amino acids bonded to each other by amide bonds, also known as peptide linkage. There are 20 different amino acids that may combine in different manner to form a diverse range of proteins. Thousands of proteins are present in a single cell and they perform numerous functions in living organisms. They are the structural components of skin, muscles, and tendons. Proteins are involved in catalysis and regulate many important reactions in our bodies, act as transport molecules that carry oxygen to most remote areas, act as enzymes and hormones, provide mechanical support, keep us healthy as part of the immune system, and transmit messages from cell to cell and control the overall growth and development.

Amino acids are organic compounds containing an amino (-NH$_2$) group and a carboxyl (-COOH) group, bonded to the same carbon. The central carbon is called α-carbon, and the other two groups which are attached to it are hydrogen atom (-H) and a variable side chain (-R). The general structure of an amino acid is shown in Figure 2.1.

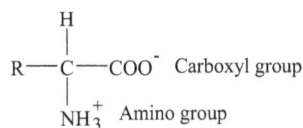

Figure 2.1: General structure of amino acid.

Although nearly about 500 naturally occurring amino acids are known, only 22 amino acids are involved in the synthesis of proteins (proteinogenic amino acids) and 20 of these proteinogenic amino acids that appear in the genetic code are known as standard amino acids (Table 2.1). Pyrrolysine and selenocysteine are two nonstandard amino acids. Amino acids can be classified on the basis of the side chain (aliphatic, aromatic, nitrogen, hydroxyl, or sulfur containing), location of core functional groups (alpha (α), beta (β), gamma (γ), or delta (δ)), polarity, and pH level. Asparagine was the first amino acid which was discovered and isolated from asparagus in 1806 by French chemists Louis-Nicolas Vauquelin and Pierre Jean Robiquet. Threonine was the last among the 20 amino acids to be discovered in 1935 by William Cumming Rose.

https://doi.org/10.1515/9783110793765-002

Table 2.1: Names and symbols of 20 standard amino acids.

S. no.	Amino acid	Three-letter abbreviation	One-letter symbol
1	Glycine	Gly	G
2	Alanine	Ala	A
3	Valine	Val	V
4	Leucine	Leu	L
5	Isoleucine	Ile	I
6	Serine	Ser	S
7	Threonine	Thr	T
8	Cysteine	Cys	C
9	Methionine	Met	M
10	Aspartate	Asp	D
11	Asparagine	Asn	N
12	Glutamate	Glu	E
13	Glutamine	Gln	Q
14	Lysine	Lys	K
15	Arginine	Arg	R
16	Histidine	His	H
17	Phenylalanine	Phe	F
18	Tyrosine	Tyr	Y
19	Tryptophan	Trp	W
20	Proline	Pro	P

2.1 Stereochemical Configuration of Amino Acids

If α-carbon is attached to all four different substituents, it is called asymmetric carbon. Except in glycine, the central carbon in all amino acids has four distinct groups and hence exhibits chirality. Molecules that exhibit chirality are also optically active, that is, they can rotate plane-polarized light. The structures of D- and L-stereoisomers of amino acids are defined relative to glyceraldehyde. Glyceraldehyde has been historically used as the standard to which chiral compounds are compared as chemical alteration had no effect on its configuration. This convention was given by Emil Fisher in 1891, and chiral compounds having configuration as D-glyceraldehyde are called D, and those having configuration as L-glyceraldehyde are termed as L. So, if the amino group is present on the right side of α-carbon, it is called D-amino acid, and if present on the left side, then it is named as L-amino acids. It is interesting to note that naturally occurring sugars belong to D-series, whereas all amino acids found in proteins are of the L-configuration (Figure 2.2). D-Amino acids seldom occur in nature but they are the constituent of bacterial cell wall. Nineteen amino acids, except proline (which has a secondary amino group), contain primary amino group (-NH_3^+).

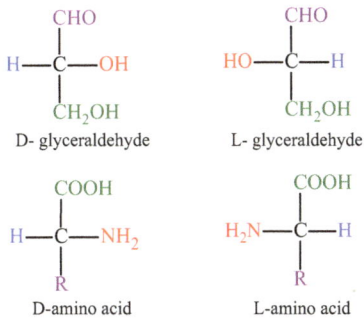

Figure 2.2: D- and L-forms of amino acid based on glyceraldehyde.

2.2 Classification of Amino Acids

Amino acids can be classified on the basis of
(i)   side chain "R,"
(ii) nutritional requirement, and
(iii) metabolic fate.

2.2.1 Classification on the Basis of Side Chain "R"

The amino acids can be classified into five main groups on the basis of their "R" group (side chain) which are as follows:

2.2.1.1 Neutral Amino Acid (nonpolar, hydrophobic side chains)

Amino acids containing aliphatic side chains belong to this category. Glycine is the simplest amino acid and has only one hydrogen atom as its side chain. Another simple amino acid in this category is alanine which contains methyl group as its side chain. Valine, leucine, and isoleucine have large hydrocarbon side chains, and methionine which has a thioether (–S–) group also belongs to this category. Isoleucine has two chiral centers as four different substituents are attached to beta(β)-carbon. Proline is unusual because it contains an "imino" group (-NH-) instead of amino group (-NH$_2$):

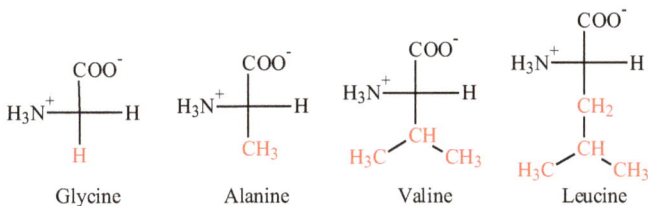

Isoleucine   Methionine   Proline

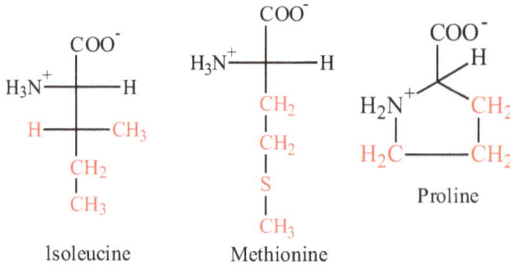

2.2.1.2 Amino Acids with Negatively Charged R Groups (Acidic Amino Acids)

Amino acids which consist of carboxyl group in their side chain, for example, aspartic acid (aspartate) and glutamic acid (glutamate) fall under this category.

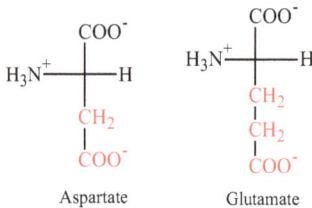

Aspartate   Glutamate

2.2.1.3 Amino Acids with Positively Charged R Groups (Basic Amino Acids)

The amino acids containing basic groups such as an amino or imidazole ring in their side chain come under this category, and are known as basic amino acids. For example, lysine, arginine, and histidine belong to the category of basic amino acids:

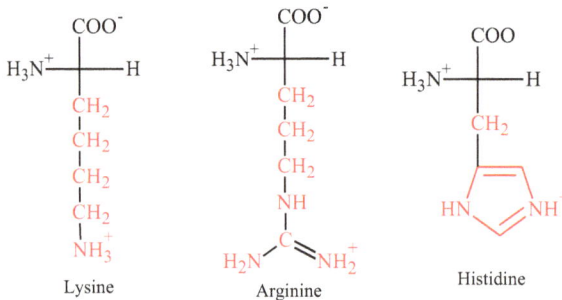

Lysine   Arginine   Histidine

2.2.1.4 Amino Acids with Polar Groups (Uncharged R Group)

These amino acids do not carry any charge in their side chain but have functional groups that can be linked with water via hydrogen bonds and hence they are soluble in water. Serine, threonine, cysteine, asparagine, and glutamine belong to this category. Serine and threonine have hydroxyl groups which provide polarity to these amino acids. Similar to isoleucine, threonine also contains an additional asymmetric center and hence can exist in four possible stereoisomers. Cysteine contains sulfhydryl (thiol, -SH) group, and on oxidation can be linked with another cysteine via a disulfide bond. The other two amino acids in this category are asparagine and glutamine which have amide (-CONH$_2$) group as their side chains:

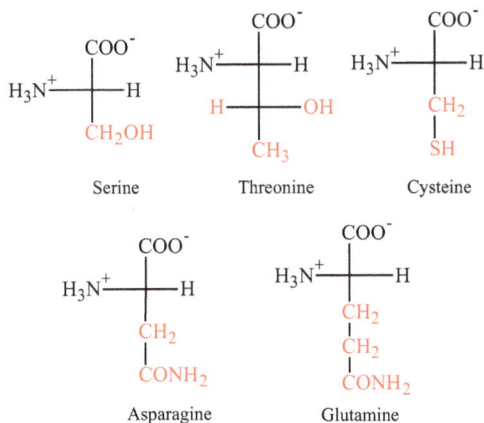

Serine   Threonine   Cysteine

Asparagine   Glutamine

2.2.1.5 Amino Acids with Aromatic Side Chain

These amino acids have aromatic side chains, and examples in this category include phenylalanine, tyrosine, and tryptophan. In phenylalanine, a phenyl group is attached in place of one of the hydrogens of methyl group of alanine, tryptophan is an indole ring containing aromatic amino acid, whereas in tyrosine side chain consists of a hydroxyl group attached to the benzene ring:

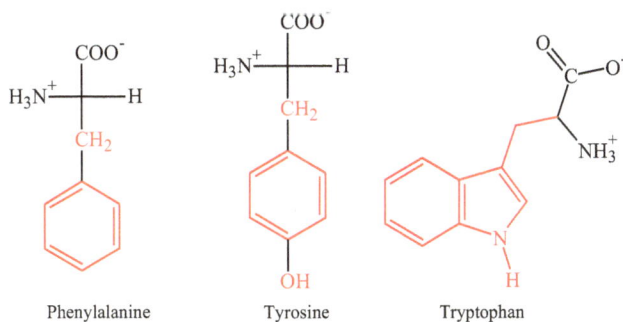

Phenylalanine   Tyrosine   Tryptophan

2.2.2 Classification on the Basis of Nutritional Requirement

Amino acids can be divided into three classes on the basis of nutritional requirement which are as follows:

1. **Essential amino acids:** These are the amino acids that cannot be synthesized in the human body and thus must be supplied through diet. They are required for the

growth and development of an individual. The amino acids which humans cannot synthesize are valine, leucine, isoleucine, methionine, phenylalanine, threonine, tryptophan, and lysine. These are also known as indispensable amino acids.

2. **Non-essential amino acids:** These are the amino acids that can be synthesized by the human body from other molecules through a chain of biochemical reactions and are not required to be taken as dietary proteins. Glycine, alanine, serine, cysteine, aspartic acid, asparagine, glutamic acid, glutamine, tyrosine, and proline are included in this category. Their functions in the body are equally important as those of essential amino acids.

3. **Semi-essential amino acids:** Semiessential amino acids are those amino acids which cannot be synthesized in the human body in sufficient amounts, for example, histidine and arginine, although these are not essential for an adult individual but are required by children, pregnant women, and lactating mothers.

2.2.3 Classification on the Basis of Metabolic Fate

Based on the type of products that are formed during the breakdown or catabolism of amino acids, they are further classified into three classes: purely glucogenic, purely ketogenic, and both glucogenic and ketogenic (Figure 2.3). The catabolism of glucogenic amino acids produces either pyruvate or one of the intermediates of the Krebs cycle, whereas catabolism of ketogenic amino acids produces acetyl-CoA (acetyl-coenzyme A) or acetoacetyl-CoA, which is a precursor of ketone bodies.

(i) **Purely glucogenic amino acids:** These amino acids serve as precursors for the synthesis of glucose via gluconeogenesis. Glycine, alanine, valine, proline, serine, aspartic acid, asparagine, glutamic acid, glutamine, methionine, cysteine, histidine, and arginine are examples of glucogenic amino acids.

(ii) **Purely ketogenic amino acids:** A ketogenic amino acid is an amino acid that can metabolized to compounds that are used to generate ketone bodies. In humans, leucine and lysine are purely ketogenic.

(iii) **Both glucogenic and ketogenic amino acids:** Amino acids such as isoleucine, phenylalanine, threonine, tryptophan, and tyrosine are classified as both ketogenic and glucogenic because some of their carbon atoms act as potential precursors of glucose whereas others appear in acetyl-CoA or acetoacetyl CoA.

2.3 Physical Properties of Amino Acids

Amino acids are nonvolatile crystalline solids which melt with decomposition at very high temperature. They are readily soluble in aqueous medium but only slightly soluble or insoluble in nonpolar organic solvents like benzene and ether. The solution of amino acids has very high dielectric constant. Although the amino acids are commonly written

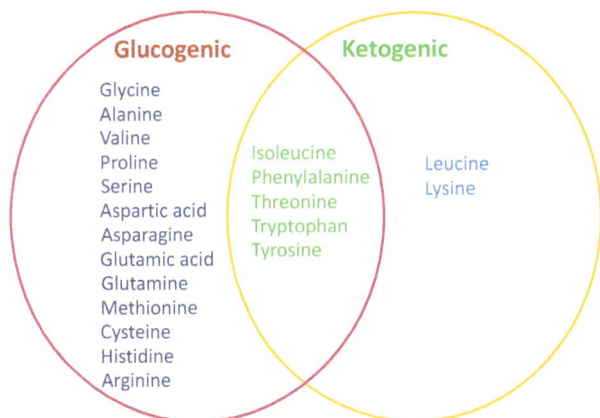

Figure 2.3: Classification of amino acids on the basis of metabolic fate.

as NH_2-CHR-COOH, but in actual, the acidic -COOH and basic -NH_2 groups react with one another to form an internal salt, called zwitterion. A zwitterion is a compound that bears a negative charge on one atom and a positive charge on a nonadjacent atom. At the isoelectric point (or isoelectric pH), an amino acid exists as a "zwitterion" or "dipolar ion." Isoelectric point or isoionic point is the pH at which an amino acid carries no electrical charge and does not migrate under electric field. Figure 2.4 shows the structure of amino acid under different solution conditions.

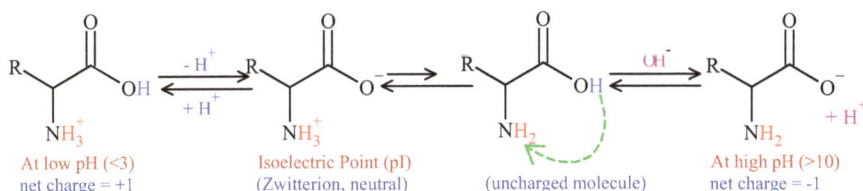

Figure 2.4: Structure of amino acid at different pH.

Amino acids are characterized by two pK_a values: pk_{a1} for the carboxylic acid and pK_{a2} for the amino group. They exist in their acidic form in solutions that are more acidic than their pK_a values and in basic forms in solutions that are more basic than their pK_a values. The carboxylic groups of amino acids have pK_a values of approximately 2.0 and the protonated amino groups have pK_a values near about 9.0. So, the pH of the solution is the deciding factor for the existence of zwitterion of any amino acid. The isoelectric point will be halfway between or the average of these two pK_a values, that is, pI for any amino acid containing neutral side chain can be calculated as follows:

$$pI = \frac{pK_{a1} + pK_{a2}}{2}$$

For the simplest amino acid, glycine, $pK_{a1} = 2.4$ and $pK_{a2} = 9.8$, so $pI = 6.1$. Glycine with a pI of 6.1 exists as a positively charged species at a pH below 6.1 and it will move toward the negative electrode. Similarly, glycine has a negative charge (–1) in solutions that have a pH above 6.1 and its movement will be toward positive electrode.

When an amino acid contains an ionizable side chain, it involves a third acid dissociation constant, pK_{a3}, and the situation is different in such cases. For amino acids containing acidic side chain (aspartic acid and glutamic acid), pI will be at the lower pH because the acidic side chain introduces an "extra" negative charge, and similarly, for basic amino acids (lysine and arginine), pI occurs at higher pH. In such cases, isoelectric point (pI value) is the average of the pK_a values of similarly ionizable groups. For example, aspartic acid, an acidic amino acid, has an extra carboxyl group on its second carbon. Its pI value can be calculated as

$$pI = \frac{pK_a(\text{COOH}) + pK_a(\text{side chain COOH})}{2}$$

For aspartic acid, $pK_{a1} = 2.1$ and $pK_{a3} = 3.9$, so $pI = 3.0$. So, it exists as neutral species at pH 3.0 and form negative ions with charges –1 and –2 at pH values greater than 3.0 as shown in Figure 2.5.

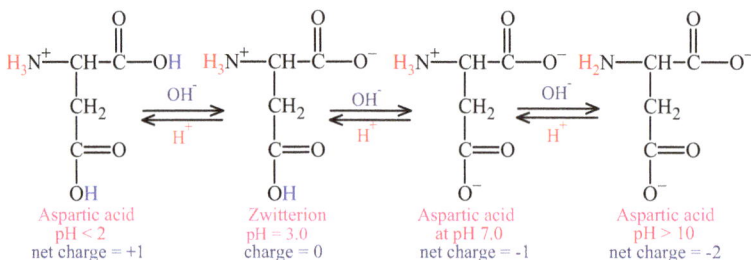

Figure 2.5: Structure of aspartic acid at different pH.

Table 2.2 summarizes pK_a values for various amino acids. It is important to mention that amino acid can never exist as uncharged compound regardless of the pH of the solution.

2.3.1 Electrophoresis

Electrophoresis is a technique which is employed for the separation of mixture of amino acids or nucleic acids on the basis of migration toward cathode and anode. Although this method was first used by Arne Tiselius in 1931 for the separation of different substances from one another; new techniques of chemical analysis and separation

Table 2.2: pK_a values of amino acids.

Amino acid	pK_{a1} (α-COOH)	pK_{a2} (α-NH$_3$$^+$)	pK_{a3} side chain (-R)
Alanine	2.4	9.9	
Arginine	1.8	9.0	12.5
Asparagine	2.1	8.8	
Aspartic acid	2.1	9.9	3.9
Cysteine	1.9	10.8	8.3
Glutamic acid	2.1	9.5	4.1
Glutamine	2.2	9.1	
Glycine	2.4	9.8	
Histidine	1.8	9.3	6.0
Isoleucine	2.3	9.8	
Leucine	2.3	9.7	
Lysine	2.2	9.2	10.8
Methionine	2.1	9.3	
Phenylalanine	2.2	9.2	
Proline	2.0	10.6	
Serine	2.2	9.2	
Threonine	2.1	9.1	
Tryptophan	2.4	9.4	
Tyrosine	2.2	9.1	10.1
Valine	2.2	9.7	

processes based on electrophoresis continued to be developed till twenty-first century. This method depends on pH of the solution and electric charge on amino acids. All the amino acids differ from each other in their isoelectric points. The principle behind this technique is migration of amino acids toward a particular electrode (cathode or anode) at a pH different from its isoelectric point. The mixture of amino acids is applied on to an inert support, which can be a paper or some polymeric gel and accordingly, it is termed as paper electrophoresis or gel electrophoresis. Here, we have described the separation of amino acids using paper electrophoresis. In this procedure, a suitable buffer solution of the desired pH is applied on a strip of filter, and the mixture of amino acids is applied at the center of the paper strip. The ends of paper strip are fixed to dip in buffer solutions in two troughs which are fitted with electrodes (cathode and anode), and an electric potential is applied at its end. On applying electric current through the strip, the charged amino acids migrate along the strip toward the respective electrodes of opposite polarity. For example, a mixture of alanine, aspartic acid, and lysine at pH 6.0, when applied on a paper strip, and on passing electric current, will show the migration of amino acids as displayed in Figure 2.6.

The maroon spot shows the original position of lysine and aspartic acid before application of electric field. On applying electric field, lysine migrated toward cathode, whereas aspartic acid migrated toward anode. Alanine has net charge zero and therefore does not show any movement, and its position is same before and after the application of electric field. Amino acids with isoelectric point (pI value) lower than the buffer

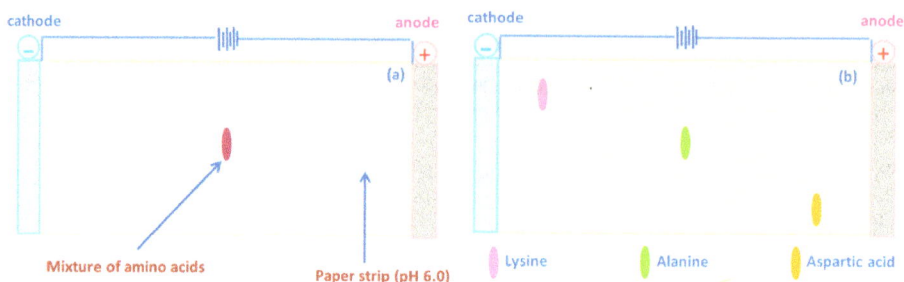

Figure 2.6: Separation of amino acids by electrophoresis (a) before separation and (b) after separation.

pH will lose a proton and acquire a negative charge (aspartic acid, pI = 3.0) and will move toward anode. On the other hand, amino acids with isoelectric point (pI value) greater than the buffer pH will gain a proton and become positively charged (lysine, pI = 10.0) and thus migrate toward cathode. Alanine has pI value same to that of the buffer pH and does not migrate toward any electrode under the influence of electric field. Thus, a mixture of amino acids can be separated and analyzed using electrophoresis.

2.4 Synthesis of Amino Acids

Amino acids can be synthesized by various methods; however, a single approach is not sufficed for the preparation of all amino acids. There are some methods which are extensively used to prepare a particular amino acid, whereas others can be employed to synthesize more than one amino acid. Some of these methods are:

1. From α-halogenated acids: Amino acids are synthesized from the reaction of α-halocarboxylic acids with ammonia. The α-halocarboxylic acid can be obtained via α-substitution reaction of a carboxylic acid by Hell–Volhard–Zelinsky (HVZ) reaction. This is an example of a nucleophilic substitution in which halide group is substituted by the $-NH_2$ group. This method is used to prepare glycine, alanine, valine, leucine, and aspartic acid:

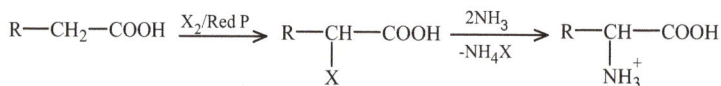

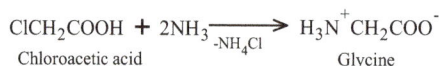

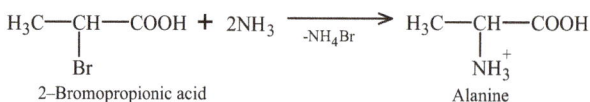

$$R-CH_2-COOH \xrightarrow{X_2/Red\ P} R-\underset{\underset{X}{|}}{C}H-COOH \xrightarrow[-NH_4X]{2NH_3} R-\underset{\underset{NH_3^+}{|}}{C}H-COOH$$

$$ClCH_2COOH + 2NH_3 \xrightarrow[-NH_4Cl]{} H_3N^+CH_2COO^-$$
$$\text{Chloroacetic acid} \qquad\qquad \text{Glycine}$$

$$H_3C-\underset{\underset{Br}{|}}{C}H-COOH + 2NH_3 \xrightarrow[-NH_4Br]{} H_3C-\underset{\underset{NH_3^+}{|}}{C}H-COOH$$
$$\text{2–Bromopropionic acid} \qquad\qquad \text{Alanine}$$

2. Gabriel's phthalimide synthesis: This method involves treatment of an ester of α-halo acids with potassium phthalimide. The halogen in α-halo esters is replaced by a phthalimido group, resulting in the formation of a substituted phthalimide, which on hydrolysis gives phthalic acid and α-amino acid. Better yields of amino acids are obtained by this method:

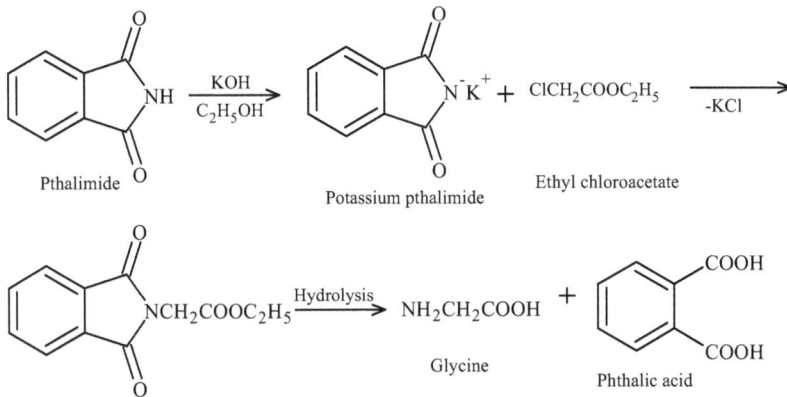

Phthalimide $\xrightarrow[C_2H_5OH]{KOH}$ Potassium phthalimide $+\ ClCH_2COOC_2H_5 \xrightarrow{-KCl}$ Ethyl chloroacetate

$NCH_2COOC_2H_5 \xrightarrow{Hydrolysis} NH_2CH_2COOH$ (Glycine) $+$ Phthalic acid (COOH, COOH)

3. From diethyl malonate: This method is basically an extension of the first method (synthesis of amino acids from α-halogenated acid) because it provides a means of preparing α-halo acids. This method is used to synthesize phenylalanine, proline, leucine, isoleucine, and methionine. Various steps involved in the synthesis of α-amino acid from diethyl malonate are shown further:

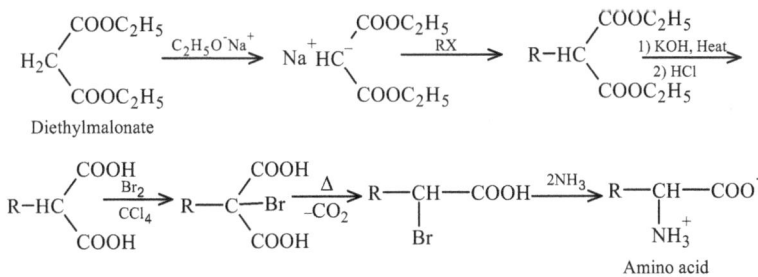

$H_2C(COOC_2H_5)_2$ (Diethylmalonate) $\xrightarrow{C_2H_5O^-Na^+} Na^+ \bar{H}C(COOC_2H_5)_2 \xrightarrow{RX} R-HC(COOC_2H_5)_2 \xrightarrow[2)\ HCl]{1)\ KOH,\ Heat}$

$R-HC(COOH)_2 \xrightarrow[CCl_4]{Br_2} R-\underset{COOH}{\overset{COOH}{C}}-Br \xrightarrow[-CO_2]{\Delta} R-\underset{Br}{CH}-COOH \xrightarrow{2NH_3} R-\underset{\overset{+}{NH_3}}{CH}-COO^-$ (Amino acid)

Phenylalanine can be synthesized starting from diethylmalonate by the following scheme:

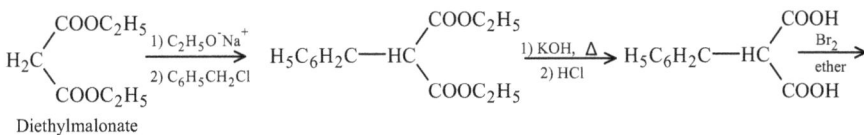

$H_2C(COOC_2H_5)_2$ (Diethylmalonate) $\xrightarrow[2)\ C_6H_5CH_2Cl]{1)\ C_2H_5O^-Na^+} H_5C_6H_2C-HC(COOC_2H_5)_2 \xrightarrow[2)\ HCl]{1)\ KOH,\ \Delta} H_5C_6H_2C-HC(COOH)_2 \xrightarrow[ether]{Br_2}$

$$H_5C_6H_2C-\underset{\underset{COOH}{|}}{\overset{\overset{COOH}{\diagup}}{C}}-Br \xrightarrow[CO_2]{\Delta} H_5C_6H_2C-\underset{\underset{Br}{|}}{CH}-COOH \xrightarrow{2NH_3} H_5C_6H_2C-\underset{\underset{NH_3^+}{|}}{CH}-COO^-$$

Phenylalanine

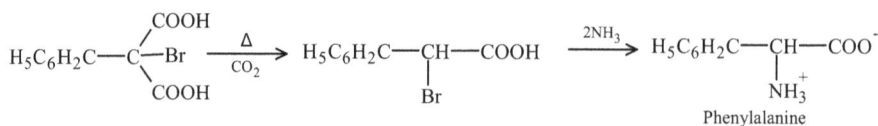

4. Strecker's synthesis: The Strecker amino acid synthesis was devised by the German chemist Adolph Strecker, and is used to synthesize an amino acid from an aldehyde or ketone using ammonium chloride and potassium cyanide. However, the original Strecker method (1850) employed ammonia and hydrogen cyanide. Later on, a safer protocol was adopted, which utilized ammonium chloride (NH_4Cl) and potassium cyanide (KCN):

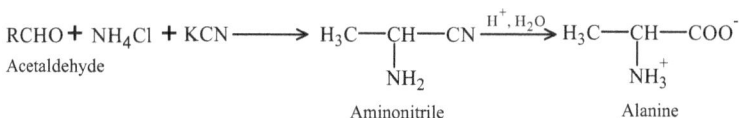

$$\underset{\text{Acetaldehyde}}{RCHO} + NH_4Cl + KCN \longrightarrow H_3C-\underset{\underset{NH_2}{|}}{CH}-CN \xrightarrow{H^+, H_2O} H_3C-\underset{\underset{NH_3^+}{|}}{CH}-COO^-$$

Aminonitrile   Alanine

In this reaction, mildly acidic ammonium ion protonates the aldehyde, increases the electrophilic character of carbon, and hence activating it toward attack with ammonia. The nucleophilic attack of ammonia to aldehyde followed by elimination of water molecule results in the formation of an aldimine. The addition of cyanide ion to aldimine forms an aminonitrile which on hydrolysis gives the desired α-amino acid. This method is widely used for the preparation of glycine, alanine, serine, valine, methionine, glutamic acid, leucine, isoleucine, and phenylalanine. The various steps involved in the reaction are summarized further.

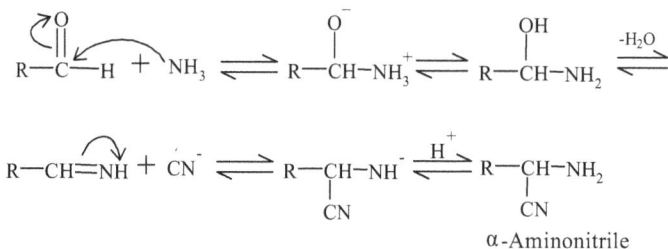

$$R-\overset{\overset{O}{\parallel}}{C}-H + NH_3 \rightleftharpoons R-\underset{\underset{NH_3^+}{|}}{\overset{\overset{O^-}{|}}{CH}} \rightleftharpoons R-\underset{\underset{NH_2}{|}}{\overset{\overset{OH}{|}}{CH}} \xrightarrow{-H_2O}$$

$$R-CH{=}NH + CN^- \rightleftharpoons R-\underset{\underset{CN}{|}}{CH}-NH^- \xrightarrow{H^+} R-\underset{\underset{CN}{|}}{CH}-NH_2$$

α-Aminonitrile

5. N-Phthalimido malonic ester synthesis: The malonic ester and Gabriel phthalimide synthesis can be combined to obtain various amino acids such as aspartic acid, glutamic acid, phenylalanine, tyrosine, cysteine, proline, serine, methionine, and lysine. N-Phthalimido malonic ester is prepared by condensing phthalimide with monobromomalonic ester, which then reacts with appropriate chloride in the presence of sodium ethoxide:

$H_2C(COOC_2H_5)_2$ Diethylmalonate

$\xrightarrow[CCl_4]{Br_2}$

Br—HC(COOC_2H_5)_2 Diethyl 2-bromopropanedioate

Potassium pthalimide + Br—HC(COOC_2H_5)_2 Diethyl 2-bromopropanedioate $\xrightarrow[-KBr]{Hydrolysis}$ N-Phthalimidomalonic ester

1) Aspartic acid

N-Phthalimidomalonic ester $\xrightarrow[2) ClCH_2COOC_2H_5]{1) C_2H_5ONa}$ $\xrightarrow[Heat]{conc. HCl}$

$\xrightarrow{-CO_2}$ $\xrightarrow{H^+, H_2O}$ Aspartic acid + Phthalic acid

2) Phenylalanine

N-Phthalimidomalonic ester $\xrightarrow[2) C_6H_5CH_2Br]{1) C_2H_5ONa}$ $\xrightarrow[Heat]{conc. HCl}$

$\xrightarrow[-CO_2]{H^+, H_2O}$ Phenylalanine + Phthalic acid

3) Cysteine

Benzylthiol —CH_2SH + HCHO + HCl ⟶ —CH_2SCH_2Cl Benzylthiomethyl chloride

N-Phthalimidomalonic ester $\xrightarrow[2) C_6H_5CH_2SCH_2Cl]{1) C_2H_5ONa}$

$\xrightarrow[2) HCl]{1) C_2H_5ONa}$ $H_5C_6H_2C-S-CH_2-\overset{NH_2}{CH}-COOH$ $\xrightarrow[Liq. NH_3]{Reduction}$ $HS-CH_2-\overset{NH_3^+}{CH}-COO^-$ Cysteine

↓ Oxidation

$HOOC-\overset{NH_2}{CH}-CH_2-S-S-CH_2-\overset{NH_2}{CH}-COOH$ Cystine

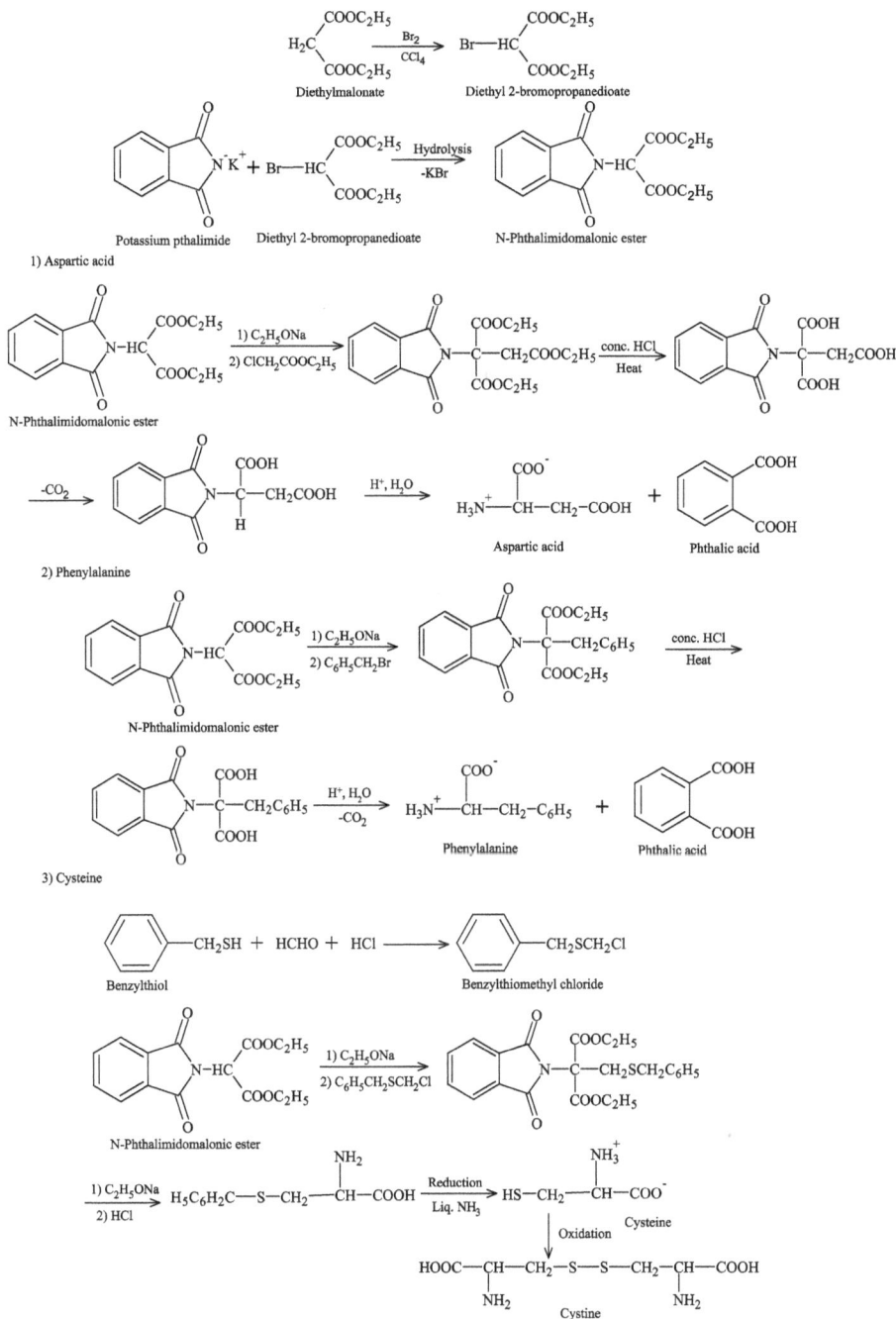

6. The azlactone synthesis (the Erlenmeyer azlactone synthesis): This method was originally introduced by Erlenmeyer for the synthesis of aromatic amino acids. In this

method, an aromatic aldehyde is heated with hippuric acid (benzoyl glycine) in the presence of acetic anhydride and sodium acetate which results in the formation of azlactone (4-benzylidene-2-phenyloxazol-5-one). The formed azlactone is then hydrolyzed using 1% sodium hydroxide solution, which opens up the ring. The intermediate is subsequently reduced using sodium amalgam and then hydrolyzed to give an amino acid, for example, various steps involved in the synthesis of tyrosine are given as follows:

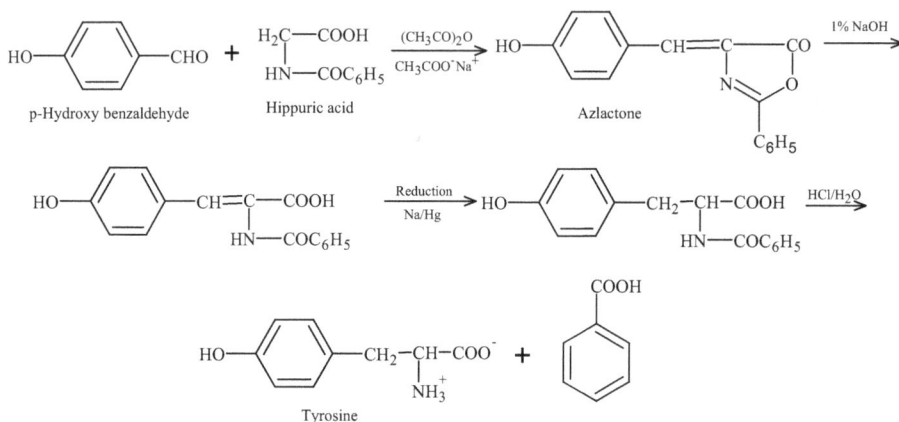

2.5 Chemical Reactions of Amino Acids

Amino acids are amphoteric in nature that is why they undergo reactions which involve both amino group and carboxyl group. The reactivity of these functional groups is particularly important to form peptides and proteins.

2.5.1 Reactions due to Amino (-NH$_2$) Group

2.5.1.1 Salt formation
Amino acids form salts with strong mineral acids:

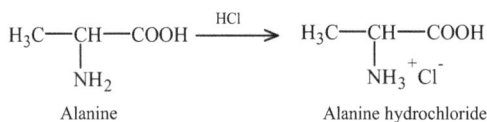

2.5.1.2 Acylation and benzoylation
Amino acids can be acetylated using acid chlorides in aqueous alkali or using acetic anhydride. For example,

$$H_2N-CH_2-COOH + (CH_3CO)_2O \longrightarrow \underset{\text{Acetyl glycine}}{H_3C-\overset{\overset{\displaystyle O}{\|}}{C}-NH-CH_2-COOH}$$

H_2N—CH_2—COOH (Glycine) + (CH_3CO)_2O (Acetic anhydride)

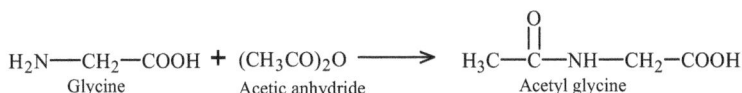

Similarly, benzoylation of amino group can be carried out with benzoyl chloride in alkaline medium to form *N*-benzoyl amino acids. This reaction is known as Schotten–Baumann reaction. However, it is necessary to add acid to the reaction mixture to obtain the product:

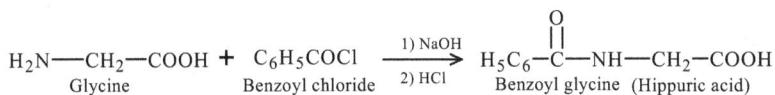

$$H_2N-CH_2-COOH + C_6H_5COCl \xrightarrow[\text{2) HCl}]{\text{1) NaOH}} \underset{\text{Benzoyl glycine (Hippuric acid)}}{H_5C_6-\overset{\overset{\displaystyle O}{\|}}{C}-NH-CH_2-COOH}$$

Glycine + Benzoyl chloride

2.5.1.3 Alkylation

Amino acids, like amines, can be alkylated using alkyl halides to get monoalkyl amino acids. However, in the presence of excess alkyl halides, formation of quaternary alkyl ammonium salts known as betaines having zwitterionic nature takes place. Thus, glycine on treatment with excess of methyl iodide gives *N*, *N*, *N*-trimethylglycine:

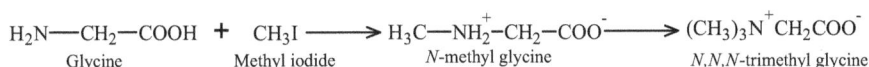

$$H_2N-CH_2-COOH + CH_3I \longrightarrow H_3C-\overset{+}{N}H_2-CH_2-COO^- \longrightarrow (CH_3)_3\overset{+}{N}CH_2COO^-$$

Glycine + Methyl iodide → *N*-methyl glycine → *N,N,N*-trimethyl glycine

2.5.1.4 Reaction with Nitrous Acid

Similar to the reaction of primary aliphatic amines, amino group of amino acids reacts with nitrous acid and undergoes deamination with the evolution of nitrogen gas to give the corresponding α-hydroxy carboxylic acid. This method is employed for gasometric determination of amino acids, known as Van Slyke method:

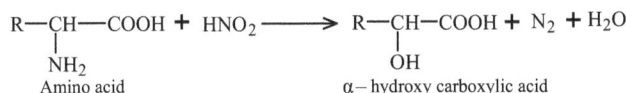

$$\underset{\underset{\text{Amino acid}}{NH_2}}{R-CH-COOH} + HNO_2 \longrightarrow \underset{\underset{\alpha-\text{hydroxy carboxylic acid}}{OH}}{R-CH-COOH} + N_2 + H_2O$$

2.5.1.5 Reaction with Formaldehyde

The amino group ($-NH_2$) of amino acids reacts with formaldehyde at room temperature to form methylene amino acids. This method is significantly known as Sorensen's formal titration method and is used for the estimation of amino acids present in a sample by direct titration with alkali. Amino acids and formaldehyde solution must be neutralized using alkali before proceeding for titration. Conversion of amino acid to methylene amino acid takes place via reaction with formaldehyde which can be easily estimated by titrating directly with alkali:

$$\underset{\text{Glycine}}{H_2NCH_2COOH} + \underset{\text{Formaldehyde}}{HCHO} \longrightarrow \underset{\text{\textit{N}-Methylene glycine}}{H_2C{=}NCH_2COOH}$$

$$\underset{\text{Glycine}}{H_2NCH_2COOH} + \underset{\text{Formaldehyde}}{2HCHO} \longrightarrow \underset{\text{Dimethylol glycine}}{(CH_2OH)_2NCH_2COOH}$$

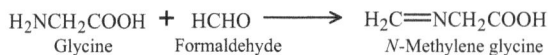

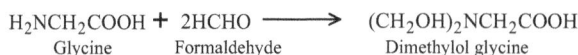

2.5.1.6 Oxidation

An amino acid can be oxidized to imino group by treating with potassium permanganate or hydrogen peroxide, which then undergoes hydrolysis, resulting in the formation of a keto acid. In biological systems, oxidation and reduction reactions are carried out by a class of enzymes, oxidoreductases:

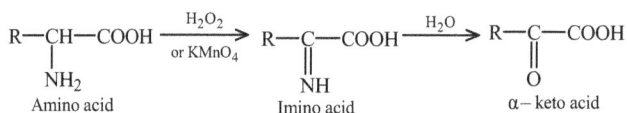

2.5.1.7 Deamination

Amino acids on heating with hydroiodic acid undergo deamination to give carboxylic acids:

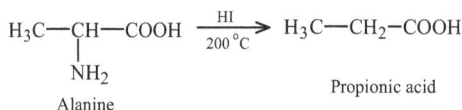

2.5.1.8 Reaction with Nitrosyl Chloride

On treatment with nitrosyl chloride (or bromide), amino acid forms chloro (or bromo) acids:

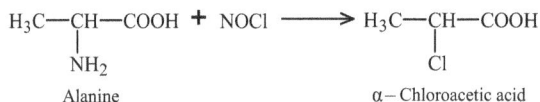

2.5.2 Reactions due to Carboxyl (-COOH) Group

2.5.2.1 Esterification

The amino acids can be esterified by refluxing with absolute alcohol in the presence of dry hydrogen chloride. The resultant product amino ester hydrochlorides on treatment with cold dilute alkali produces free amino acid ester:

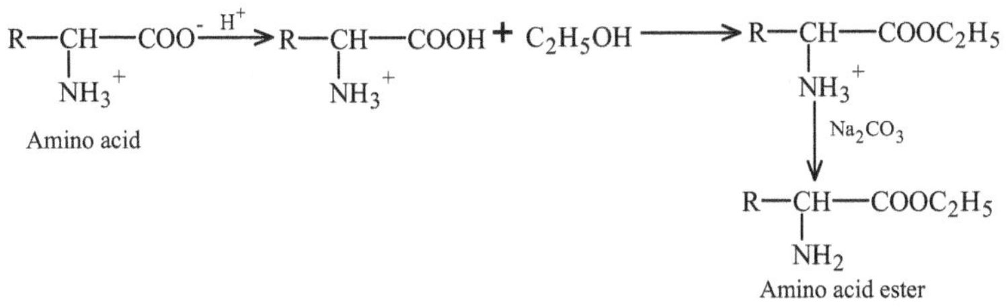

$$R-\underset{\underset{NH_3^+}{|}}{CH}-COO^- \xrightarrow{H^+} R-\underset{\underset{NH_3^+}{|}}{CH}-COOH + C_2H_5OH \longrightarrow R-\underset{\underset{NH_3^+}{|}}{CH}-COOC_2H_5$$

Amino acid

$$\downarrow Na_2CO_3$$

$$R-\underset{\underset{NH_2}{|}}{CH}-COOC_2H_5$$

Amino acid ester

2.5.2.2 Decarboxylation

Amino acids undergo decarboxylation in the presence of barium hydroxide to form amines:

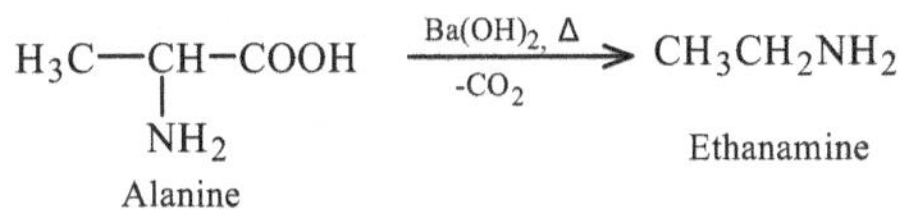

$$H_3C-\underset{\underset{NH_2}{|}}{CH}-COOH \xrightarrow[-CO_2]{Ba(OH)_2,\ \Delta} CH_3CH_2NH_2$$

Alanine   Ethanamine

2.5.2.3 Reduction

On heating with lithium aluminum hydride, amino acids are reduced to the corresponding alcohols:

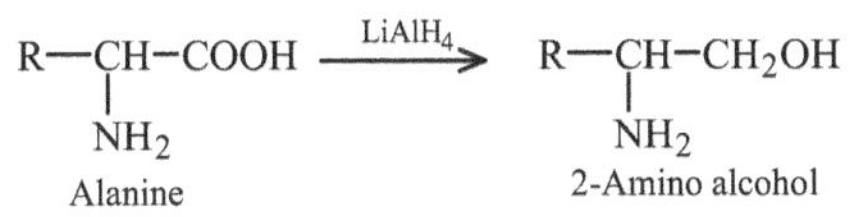

$$R-\underset{\underset{NH_2}{|}}{CH}-COOH \xrightarrow{LiAlH_4} R-\underset{\underset{NH_2}{|}}{CH}-CH_2OH$$

Alanine   2-Amino alcohol

2.5.3 Reactions Involving Both Amino and Carboxyl Groups

2.5.3.1 Action of heat

When two molecules of amino acids are heated, they undergo intramolecular dehydration and form 2,5-diketopiperazine:

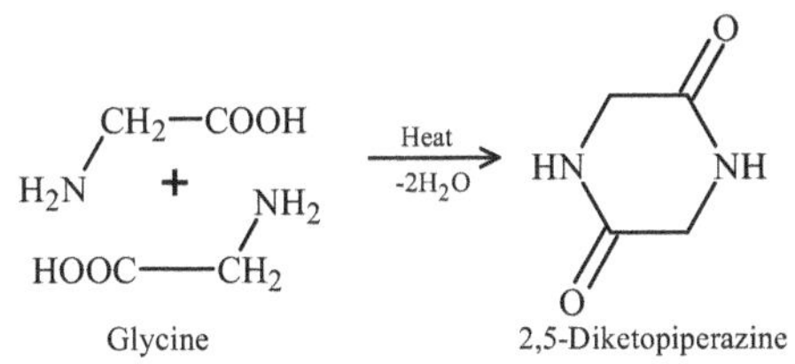

Glycine   2,5-Diketopiperazine

2.5.3.2 Formation of Metal Chelates

α-Amino acid, like other carboxylic acids, forms metal salts. For example, glycine on reaction with cupric oxide in water produces deep blue color complex:

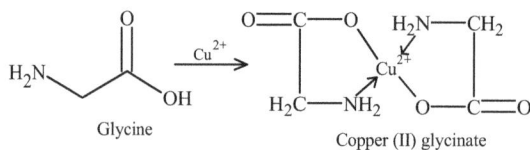

Glycine

Copper (II) glycinate

2.5.3.3 Reaction with Ninhydrin

Amino acids react with ninhydrin (2, 2-dihydroxyindane-1, 3-dione) to give a purple-colored product. This reaction is used for qualitative (in chromatography techniques) as well as for quantitative estimation of α-amino acids. Proline, a secondary amine, gives a yellow-colored product:

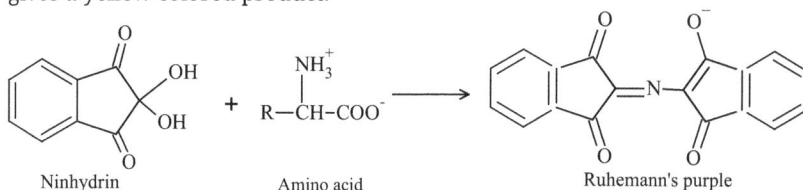

Ninhydrin

Amino acid

Ruhemann's purple

2.5.3.4 Reaction with Isocyanates

Amino acids on treatment with isocyanates form substituted urea derivatives (carbamides) which on reaction with hydrochloric acid and upon heating form hydantoins:

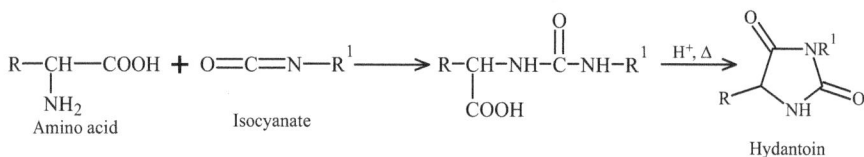

Amino acid

Isocyanate

Hydantoin

2.6 Structure of Polypeptides

Peptides are the polymers which are made up of monomers called amino acids. The size of peptide ranges from small chains consisting of few amino acids to large chains consisting of many amino acids. When two amino acids combine via an intermolecular reaction, carboxyl group of one amino acid condenses with amino group of another amino acid and forms an amide bond. The amide bond between amino acids is called a peptide linkage and the compounds formed are called peptides. Peptides are named starting with the N-terminal amino acid. The names of amino acid residues involved in peptide linkages are given suffix "yl" (e.g., glycine to glycyl) for all amino acids except the C-terminal acids. For example, a dipeptide formed from one molecule of glycine and one molecule of alanine is named as glycylalanine:

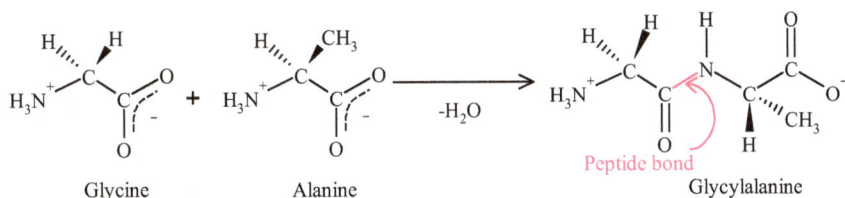

Glycine   Alanine   Glycylalanine

Peptides can be classified as dipeptides, tripeptides, or tetrapeptides based on the number of amino acid residues in it. A peptide derived from two molecules of same or different amino acid is known as dipeptide. A large number of amino acids linked via peptide bonds result in a polypeptide chain, and each amino acid in a polypeptide chain is called a residue since that is the portion which is left after the loss of water molecule in a dehydration process. A polypeptide chain has a directionality because it has two distinct ends. A free amino group present at one end is called N-terminal, and an α-carboxyl group at the other end is called C-terminal. By convention, the sequence of amino acids in a polypeptide chain is written with N-terminal residue on the left-hand side and C-terminal residue is written on the right-hand side of the chain. For example, in the polypeptide chain Ala-Tyr-Cys-Gly-Ser (AYCGS), alanine is the amino terminal (N-terminal) and serine is the carboxyl terminal (C-terminal) residue, and Ser-Gly-Cys-Tyr-Ala (SGCYA) is a different pentapeptide with distinct chemical properties. The structure of a tripeptide Gly–Ala–Phe (glycylalanylphenylalanine) formed from glycine, alanine, and phenylalanine is shown:

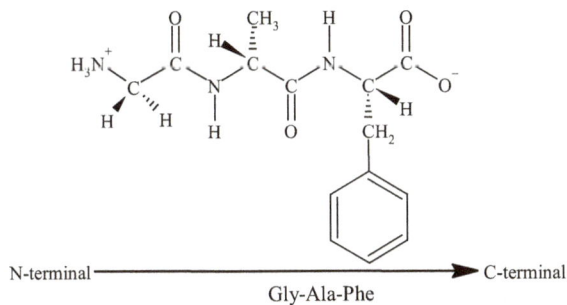

Gly-Ala-Phe

A polypeptide chain comprises two parts: a repeating part called the main chain and a variable part consisting of distinctive side chains. Polypeptide chains which consist of smaller number of amino acids are known as oligopeptides, whereas natural polypeptide chains containing 50–2,000 amino acid residues are commonly known as proteins. The mass of a protein is expressed in units of daltons (Da) and 1 Da is equal to 1 atomic mass unit. A protein having a molecular weight of 50,000 g/mol has a mass of 50,000 Da (50 kDa). *Titin*, a muscle protein, is the largest protein in the body composed of more than 34,000 amino acids and has a mass of 3 million Dalton (mDa).

Peptides play an important role in various physiological processes and some peptides also display biological activities. For example, insulin, 51 amino acid residues, consists of 2 disulfide-linked peptide chains. It is an anabolic hormone that helps the glucose to get into the cells to be used as energy, and maintains glucose level in the bloodstream within normal levels. Other examples are oxytocin and vasopressin. Oxytocin is a nonapeptide (with a disulfide linkage) which causes contraction of uterus. Vasopressin, also called antidiuretic hormone, also consists of nine amino acids which play a key role in maintaining plasma osmolality and volume of water in extracellular fluids.

2.7 Synthesis of Peptides

The synthesis of a polypeptide chain involves the step-by-step condensation of amino group of one amino acid with carboxyl group of another amino acid resulting in the formation of a peptide bond. The process of synthesis is quite difficult due to the presence of both amino and carboxyl groups in the same amino acid. There lies a possibility that the reaction may occur between the amino group and carboxyl group of same amino acids resulting in the formation of undesirable products. For example, when a simple dipeptide is synthesized from glycine and alanine, the following possibilities may be observed:

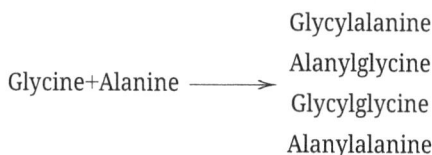

$$\text{Glycine+Alanine} \longrightarrow \begin{array}{l} \text{Glycylalanine} \\ \text{Alanylglycine} \\ \text{Glycylglycine} \\ \text{Alanylalanine} \end{array}$$

The two amino acids may condense in different ways to give a mixture of dipeptides. So, one must ensure that the amino acids that make up the polypeptide chain are in the correct order and they do not undergo any other reactions leading to the formation of undesirable products and hence affecting the overall yield of the desired peptide. Moreover, other nucleophilic groups such as -OH and -SH are also present on the side chains of amino acids which may react with the carboxylic acids. Therefore, these groups must be protected to get the desired product.

The first step in the peptide synthesis is the protection of amino group of N-terminal amino acid (first amino acid), and the second step involves the activation of carboxyl group of same amino acid. This is then followed by the formation of peptide bond with the next amino acid under the conditions that do not affect the peptide bonds or any other functional groups present in the molecule. After obtaining the desired peptide, the last step involves the removal of protecting group using a reagent that does not degrade the peptide bonds. Another way of synthesis is to protect the amino group of one amino acid and carboxyl group of another amino acid and then condensing them together in

the presence of a dehydrating agent. The general principle of polypeptide synthesis consists of the following four steps:
1. Protection of amino group
2. Protection of carboxyl group
3. Coupling reaction (formation of a peptide bond)
4. Deprotection

1. Protection of amino (-NH₂) group

Various groups have been used for the protection of an amino group. To facilitate peptide formation with minimal side reactions, some chemical groups must be introduced that can bind to the amino group and protect it from nonspecific reaction. These protecting groups must be carefully chosen so that they can be removed easily under mild conditions without affecting the newly formed peptide bond. Moreover, it should not result in the rearrangement or racemization of the neighboring chiral atom. Some important amino-protecting groups are benzyloxycarbonyl (Cbz), t-butyloxycarbonyl (Boc group), trityl (triphenyl methyl), phthaloyl, and tosyl (Ts, p-toluenesulfonyl) groups.

2. Protection of carboxyl (-COOH) group

Carboxyl groups can be protected by esterification, and specific esters like methyl, ethyl, benzyl, and t-butyl (tBu) are used for esterification. The reactive side chains of amino acids such as thiol (-SH) and hydroxyl (-OH) groups can also be protected by the benzyl group. Hydroxyl group is also protected by acetylation. The benzyl group is removed by reduction with excess sodium in liquid ammonia.

3. Coupling reaction (formation of a peptide bond)

The most important step in peptide synthesis is coupling reaction. Once the amino group and carboxyl group of amino acids are protected, coupling reaction (peptide bond formation) is carried out by converting carboxyl group of N-terminal-protected amino acid to its acyl derivative. Various methods are available for the activation of carboxyl group, and this can be done by converting the acid into acid chloride, acid azide, or p-nitrophenyl ester. The most widely used method is to use dehydrating agents such as dicyclohexylcarbodiimide (DCC) in organic solvents (methylene dichloride and tetrahydrofuran). The dipeptide ester is hydrolyzed, and the formed acid group is further activated, and reacted with an amino acid to form a tripeptide, and so on:

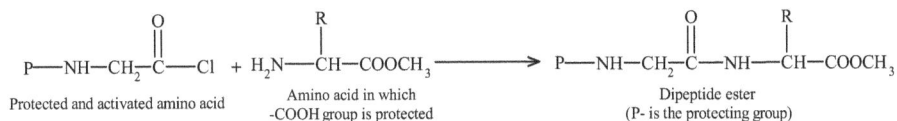

Protected and activated amino acid   Amino acid in which -COOH group is protected   Dipeptide ester (P- is the protecting group)

4. Deprotection

This is the final step in peptide synthesis, which includes removal of the protecting group. The protecting group, carbobenzyloxy, is removed by catalytic reduction but if an amino acid contains sulfur, catalytic reduction cannot be used because sulfur poisons the catalyst. In such cases, carbobenzyloxy group is removed either by sodium metal in liquid ammonia at 40 °C or by triethylsilane and palladium chloride. Hydrazine is used for the removal of N-phthaloyl group. Tosyl group is removed by hydroiodic acid at 50–60 °C, whereas acetic acid is used for the removal of trityl group. Benzyl group is removed by sodium and liquid ammonia.

2.7.1 Protection of Amino Group

Some important protecting groups for amino group of amino acids are as follows:

i. Toluene-p-sulfonyl group (tosyl)

Tosyl group was used for the protection of amino groups at the early stages for the synthesis of peptides. This group is useful for the protection of amino group of side chain of basic amino acids such as lysine and arginine. It can be easily removed by using sodium in liquid ammonia:

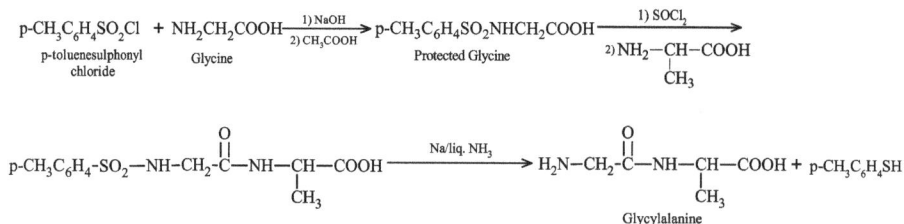

p-CH$_3$C$_6$H$_4$SO$_2$Cl + NH$_2$CH$_2$COOH $\xrightarrow[\text{2) CH}_3\text{COOH}]{\text{1) NaOH}}$ p-CH$_3$C$_6$H$_4$SO$_2$NHCH$_2$COOH $\xrightarrow[\text{2) NH}_2-\text{CH}-\text{COOH}]{\text{1) SOCl}_2}$

p-toluenesulphonyl chloride · Glycine · Protected Glycine · CH$_3$

p-CH$_3$C$_6$H$_4$-SO$_2$—NH–CH$_2$–C(=O)—NH—CH—COOH (CH$_3$) $\xrightarrow{\text{Na/liq. NH}_3}$ H$_2$N–CH$_2$–C(=O)—NH—CH—COOH (CH$_3$) + p-CH$_3$C$_6$H$_4$SH

Glycylalanine

ii. Benzyloxycarbonyl group (Cbz group)

This method was introduced by Bergmann and Zervas in 1932. They introduced the first reversible amino-protecting group for peptide synthesis, the Cbz group. The group has several advantages such as it can be (i) easily introduced into the functional group, (ii) stable to wide range of reaction conditions, and (iii) easily removed at the end of the synthesis. This group is introduced through phenylmethyl chloroformate (benzylchloroformate, C$_6$H$_5$CH$_2$OCOCl)), which is obtained readily by the action of carbonyl chloride (phosgene) on benzyl alcohol in toluene solution. This group can be condensed with the amino acid by the Schotten–Baumann reaction. The deprotection of Cbz group is carried out by catalytic hydrogenation in the presence of palladium or by using cold HBr in acetic acid. This process can be repeated multiple times to get a desired polypeptide chain:

H_5C_6-CH_2-OH + Cl—C(=O)Cl ⟶ H_5C_6-CH_2-O—COCl

Benzyl Alcohol Carbonyl Chloride Benzyl Chloroformate

H_5C_6-CH_2-O—COCl + H_2N—CH_2-COOH ⟶ H_5C_6-CH_2-O-C(=O)—NH—CH_2-COOH

Benzyl Chloroformate Glycine

1) PCl_5
2) H_2N—CH(CH_3)—COOH

H_5C_6-CH_2-O-C(=O)—NH—CH_2-CO—NH—CH(CH_3)—COOH

H_2/Pd

H_2N—CH_2-C(=O)—NH—CH(CH_3)—COOH + H_5C_6-CH_3 + CO_2

Glycylalanine (a dipeptide)

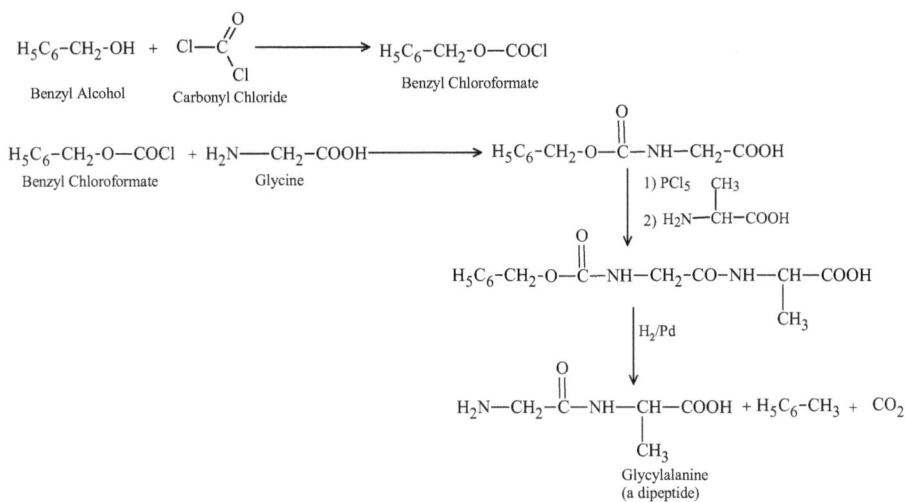

iii. Phthaloyl group

This method involves the protection of amino group using phthaloyl group. The direct reaction of phthalic anhydride and amino acids in the fused state or in organic solvents leads to partial or complete racemization. Therefore, this group can be introduced by condensing N-ethoxy carbonyl phthalimide with amino acids under basic conditions. N-Phthaloyl group can be removed by refluxing with hydrazine:

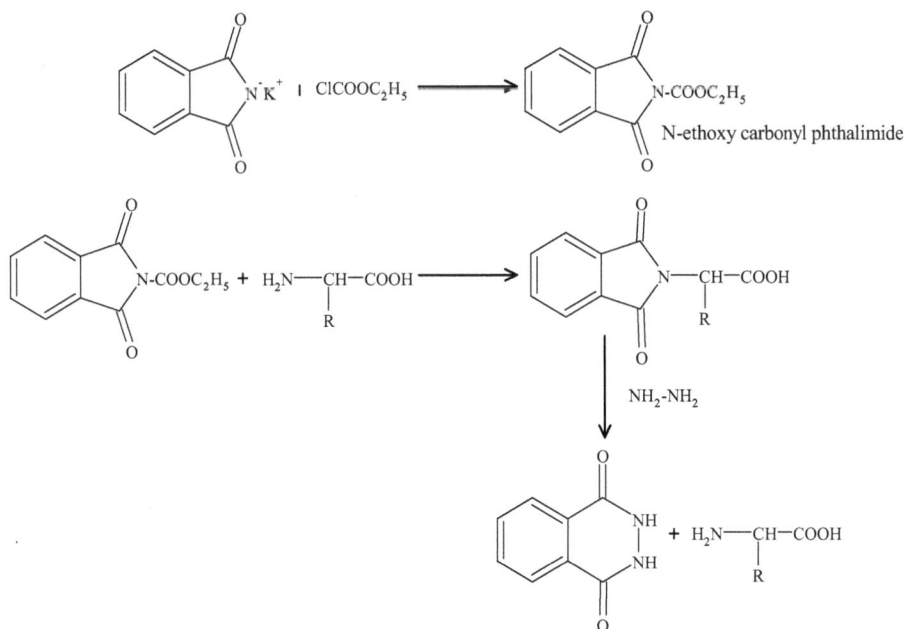

Phthalimide N^-K^+ + ClCOOC_2H_5 ⟶ N-COOC_2H_5 (N-ethoxy carbonyl phthalimide)

N-COOC_2H_5 + H_2N—CH(R)—COOH ⟶ N—CH(R)—COOH

NH_2-NH_2

phthalhydrazide (NH—NH) + H_2N—CH(R)—COOH

iv. Dimedone (5,5-dimethylcyclohexane-1,3-dione)

The use of dimedone as a protecting agent for amino groups was given by Halpern and James. In this method, dimedone (5, 5-dimethylcyclohexane-1,3-dione) reacts with amino acid esters (thiophenyl ester) to form dimedone enamine derivatives. The protecting group can be easily removed using aqueous bromine resulting in the formation of the desired peptide and 2,2-dibromodimedone:

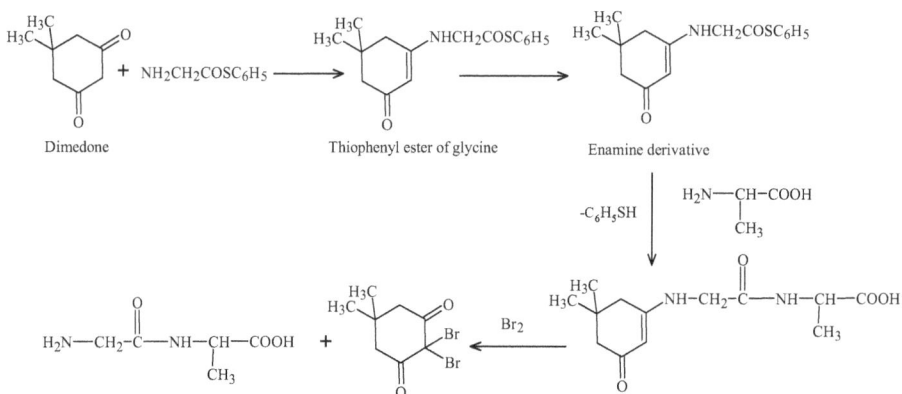

Dimedone   Thiophenyl ester of glycine   Enamine derivative

v. t-Butyloxycarbonyl group (Boc group)

This is the most commonly and widely used group employed for protecting α-amino groups. This group is named as Boc and is generally introduced into an amino acid by reaction of amino acid with bis-1,1-dimethylethyl dicarbonate (di-tert-butyldicarbonate). Boc group can be removed by treating the protected amino acid with trifluoroacetic acid in dichloromethane or 1 M trimethylsilyl chloride (TMS-Cl)-phenol in dichloromethane or dry hydrogen chloride in ether:

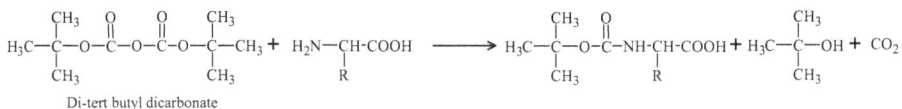

Di-tert butyl dicarbonate

vi. 2-(4-Biphenyl) isopropoxycarbonyl group (Bpoc)

This group is introduced via the azide and it forms a more stable intermediate carbocation; hence, it is preferred over Boc group. Bpoc-amino acid is used in combination with tert-butyl (tBu) side chain protection. The most common removal conditions for Bpoc group are 0.2–0.5% trifluoroacetic acid:

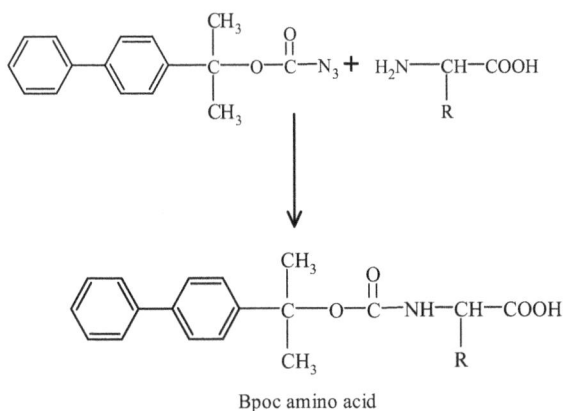

Bpoc amino acid

vii. Triphenylmethyl group (trityl or Trt)

Trityl group can be introduced by treating amino acid methyl ester with (triphenyl methyl chloride) tritylchloride followed by alkaline hydrolysis. Since trityl group is stable to bases, it is removed by mild acids such as 1% trifluoroacetic acid or 3% trichloroacetic acid:

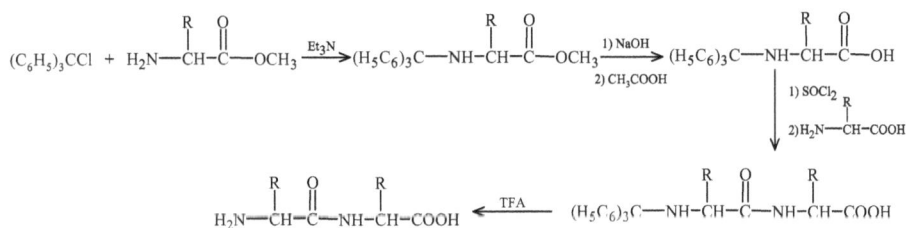

2.7.2 Protection of Carboxyl Group

The protection of C-terminal carboxyl group is different in solution and solid phase. In solid-phase peptide synthesis (SPPS), the C-terminal is usually linked to the solid support and therefore the linker acts as a protecting group. The carboxyl group is generally protected by methyl, ethyl, or benzyl alcohol in acidic conditions (HCl is generally used). The ester hydrochlorides are obtained on passing HCl gas into the suspension of amino acids in alcohol. Amino acids can also be esterified by first converting them into acid chlorides using thionyl chlorides and then reacting with an alcohol to get the ester. The ester group can be easily removed using mild sodium hydroxide:

(1) H_2N—CH—COOH $\xrightarrow{C_6H_5CH_2OH/HCl}$ H_2N—CH—$\overset{\overset{\displaystyle O}{\|}}{C}$—$OCH_2$-$C_6H_5$
          | |
          R   R

(2) H_2N—CH—COOH $\xrightarrow{SOCl_2}$ H_2N—CH—$\overset{\overset{\displaystyle O}{\|}}{C}$—Cl
          | |
          R   R

$\downarrow C_6H_5CH_2OH$

H_2N—CH—$\overset{\overset{\displaystyle O}{\|}}{C}$—$OCH_2$-$C_6H_5$
          |
          R

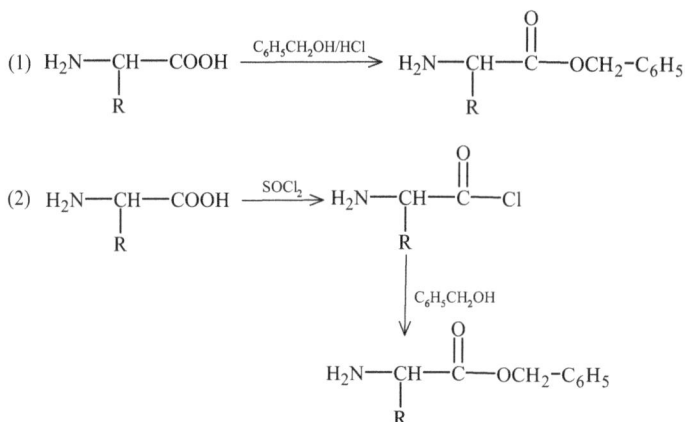

2.7.3 Protection of the Side Chain of Amino Acids

The choice of protecting groups depends on the nature of the side chain of amino acids; therefore, it is not possible to categorize them in a specific category. The protection of side chain in lysine is essential in peptide synthesis to prevent acylation. It is necessary to employ separate protecting groups for α- and ε-amino groups because the protecting group on the ε-position must remain intact until the end of the peptide synthesis process, whereas the protecting group on the α-position has to undergo deprotection for coupling with next amino acids. The most commonly used protecting group for lysine is 2-chlorobenzyloxycarbonyl (Cbz-Cl). It is generally preferred over Cbz group because Cbz-Cl is resistant to the repetitive trifluoroacetic acid treatments. Cbz-Cl is removed with hydrofluoric acid or trifluoromethane sulfonic acid. It can also be removed by hydrogenolysis in solution.

The β-carboxyl group of aspartic acid and γ-carboxyl group of glutamic acid are usually protected via acid-catalyzed esterification by benzyl or tBu groups. However, it is important to choose a protecting group for amino group, which can be easily removed using reagents that will not affect the ester groups in the side chains of aspartic or glutamic acid. Benzyl group can be removed with hydrofluoric acid or trifluoromethane sulfonic acid, whereas 90% trifluoroacetic acid in dichloromethane is employed for the removal of tBu group.

The hydroxyl group in the side chains (serine and threonine) is protected as benzyl and tBu ethers. Hydrofluoric acid is used for the removal of benzyl group, and tBu group is removed by trifluoroacetic acid. The nucleophilic thiol "-SH" group in cysteine can be alkylated, acylated, or oxidized to disulfide by aerial oxidation; hence, it is mandatory to protect the side chain of cysteine. The S-benzyl groups are most commonly used which can be introduced by dissolving cysteine in liquid ammonia, adding sufficient sodium to convert the thiol into anion and then adding benzyl chloride.

The benzyl group can be removed either by treating with hydrogen fluoride at 25 °C or reduction with sodium in liquid ammonia.

The guanidino group in the side chain of arginine is basic, and it remains protonated during peptide synthesis.

2.7.4 Coupling Methods

2.7.4.1 Curtius Azide Method

In this method, N-protected amino acid reacts with hydrazine to form the corresponding azide by reaction with nitrous acid. N-Protected amino acid azide is then condensed with amino acid ester to give the desired peptide. Various steps involved in this synthesis are:

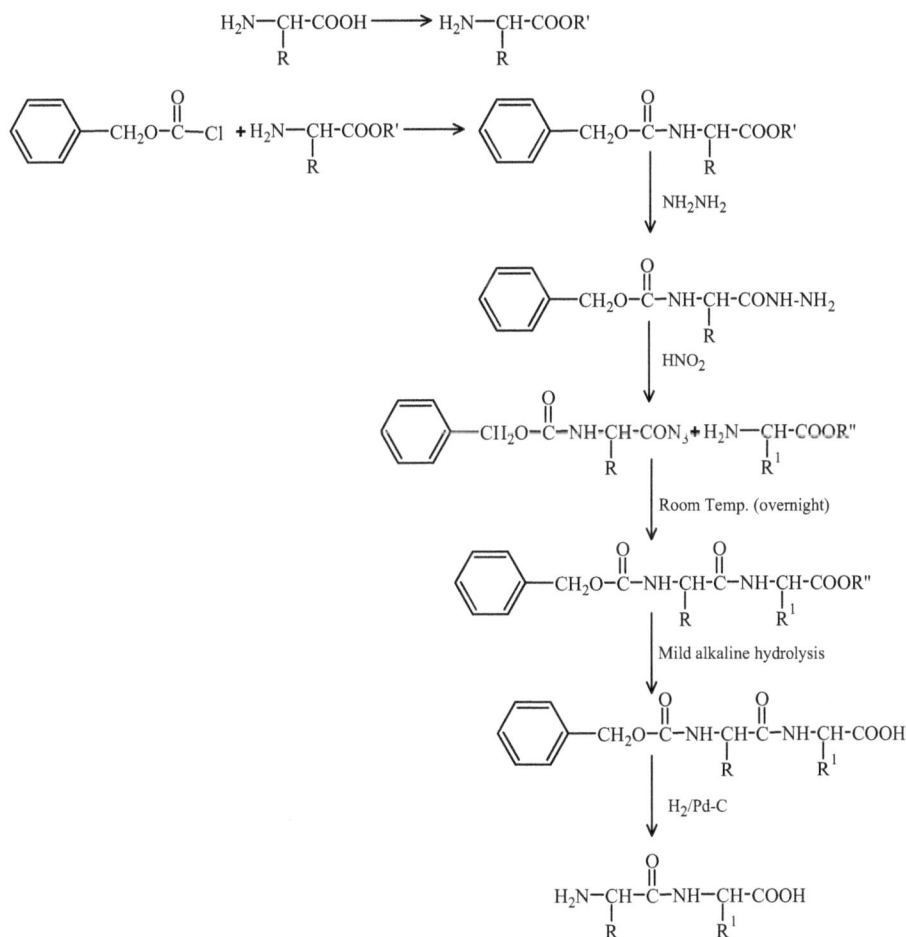

2.7.4.2 DCC Method

N, N-Dicyclohexylcarbodiimide (DCC), used as a coupling reagent in peptide synthesis for the preparation of amide bonds, was developed by Sheehan and Hess in 1955. In this method, N-carbobenzyloxy-protected amino acid reacts with an amino acid ester in the presence of DCC to give carbobenzyloxy derivative of a dipeptide ester. Various inert solvents such as methylene chloride or tetrahydrofuran are commonly used in this reaction. DCC dehydrates to form dicyclohexylurea (DCU) which separates out as a solid and can be easily filtered. In the last, the desired dipeptide is obtained by catalytic hydrogenolysis followed by hydrolysis:

N-Carbobenzyloxy protected amino acid Amino acid ester

DCC

Carbobenzyloxy derivative of adipeptide ester N, N'-Dicyclohexylurea

2.8 Solid-Phase Peptide Synthesis

Solid-Phase Peptide Synthesis (SPPS), developed by Robert B. Merrifield in 1963, is the most versatile procedure used for the synthesis of peptides. Peptides are biologically synthesized from the amino group (N-terminal) to the carboxyl group (C-terminal) in the cells, whereas in SPPS, a polypeptide chain is synthesized in the opposite direction, that is, from C-terminal to N-terminal. Merrifield used this technique in 1969 to synthesize ribonuclease, the first enzyme prepared synthetically with 124 amino acids in just 6 weeks. In this method, an amino acid is added into a peptide chain of a desired sequence while one end of the chain is attached to an insoluble synthetic resin. When the desired sequence of amino acids (a polypeptide chain) has been synthesized, a reagent can be applied to liberate it from the solid support. This method has been automated, that is, each addition of the appropriate amino acid is carried out automatically at the predetermined time. The main advantages of Merrifield synthesis are:

(i)   The ease of operation, and production of peptides are in excellent yield.

(ii) Since the solid support is insoluble, purification of the product is not required. Excess of reagents can be easily removed by simple washing with suitable solvents.

(iii) The time has been considerably reduced for the synthesis of peptides and proteins.

Merrifield solid-phase synthesis utilizes a solid support or resin of copolymer of styrene and 2% divinyl benzene. The polymer consists of long alkyl chains having a phenyl ring on every second carbon atom. It is a chloromethylated resin resulting in the attachment of the -CH$_2$Cl group.

The synthesis of a desired polypeptide chain using solid phase can be carried out in the following steps:

Step I: The N-terminal Boc--protected amino acid is attached to the copolymer via ester bond formation between -COOH group of amino acid and -CH$_2$Cl of the copolymer. Side chain functional groups must also be blocked using appropriate protecting groups.

Step II: The copolymer is washed to remove excess reagents. The Boc-protecting group is selectively removed with 50% trifluoroacetic acid in dichloromethane. The resulting α-amine salt is converted into free amino group by the addition of a tertiary amine such as diisopropylethylamine.

Step III: Another Boc-amino acid is then added to copolymer in the presence a coupling agent. The simplest and most often used method employed is coupling using DCC, which gets converted into DCU. The free amino group of copolymer-bound amino acid is then ready to couple with a second Boc-amino acid.

Step IV: The dipeptide and copolymer are again washed to remove excess reagents for the addition of next amino acid. The process is then repeated with the third N-protected amino acid and so on. The deprotection, neutralization, and coupling steps are repeated a number of times to extend the peptide chain.

Step V: Once the desired polypeptide chain is obtained, the ester bond linking it to the resin is cleaved and this is achieved by treatment with a strong anhydrous acid such as hydrofluoric acid. Various steps involved in the SPPS of a tripeptide are shown in Figure 2.7.

2.9 Determination of Sequence of Amino Acids in a Polypeptide Chain (Primary Structure)

The primary structure of a protein refers to the sequence of amino acids in a polypeptide chain. Since proteins are involved in various important biological processes, it is important to know how to determine its primary structure, that is, the number and kind of amino acids, and the order in which these amino acids are arranged in a polypeptide chain. The following steps are involved in the determination of primary structure of a polypeptide chain:

1. The first step is the isolation and purification. The purification of a polypeptide chain can be performed using various techniques such as electrophoresis, gel filtration chromatography, ion-exchange chromatography, and dialysis. Gel electrophoresis is

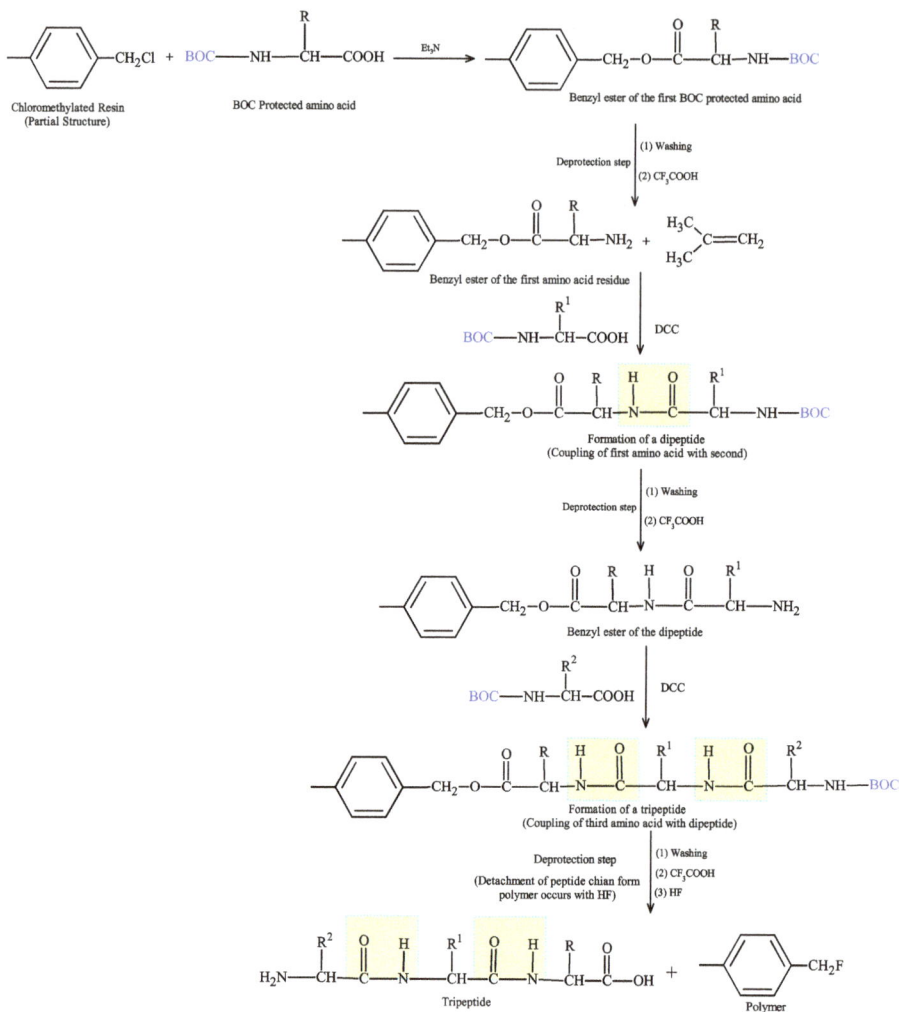

Figure 2.7: Solid-phase peptide synthesis (schematic representation for the synthesis of a tripeptide).

a powerful and widely used technique employed for the separation and purification of biomolecules like nucleic acids and proteins.

2. Then, the peptide chain is converted into its amino acid constituents by complete hydrolysis. This can be carried out under acidic or basic conditions or using enzymes. However, enzymatic hydrolysis is highly specific and only specific peptide linkages are cleaved. This step is usually carried out in the presence of an acid (6 N HCl), which results in the degradation of polypeptide chain into amino acids. Alkaline hydrolysis can also be carried out by treating the peptide with 2–4 N NaOH at 100 °C for 1–10 h. But this method is not preferred as alkaline hydrolysis can destroy some amino acids such as arginine, cysteine, serine, and threonine.

3. The next step is the qualitative and quantitative analyses of amino acid, which can be performed using ion-exchange chromatography.
4. After determining the total number and type of amino acids in a polypeptide chain, the next step is to cleave disulfide linkage which can be done using performic acid. Disulfide bridges can also be cleaved using 2-mercaptoethanol. If there is no disulfide linkage, then there is no need to perform this step.
5. The last two steps in this process are the identification of N- and C-terminals, known as end group analysis.

2.9.1 End Group Analysis

The N-terminal amino acid contains a free amino group, whereas a free carboxyl group is present at the C-terminal. By this method, the nature of N- and C-terminal groups is determined.

2.9.1.1 Amino-Terminal (N-Terminal) Analysis

Amino acids present at the N-terminal can be determined by the following three methods:

i. DNP method (Sanger's method)

The method was developed by Frederick Sanger, and in this method, the polypeptide chain under analysis is treated with 1-fluoro-2,4-dinitrobenzene (FDNB) in presence of sodium bicarbonate solution at room temperature. FDNB undergoes nucleophilic substitution by the free amino acid of polypeptide chain to form a 2,4-dinitrophenyl (DNP) derivative. The product undergoes acid hydrolysis (the cleavage of the peptide bond connecting the N-terminal amino acid to the rest of the polypeptide molecule takes place), which results in the separation of dinitrophenyl derivative of the N-terminal amino acid and the remaining amino acid residues:

FDNB

Polypeptide chain

-HF

2,4-Dinitrophenyl derivative

H^+/H_2O

2,4-DNP derivative of N-terminal amino acid

Mixture of amino acids

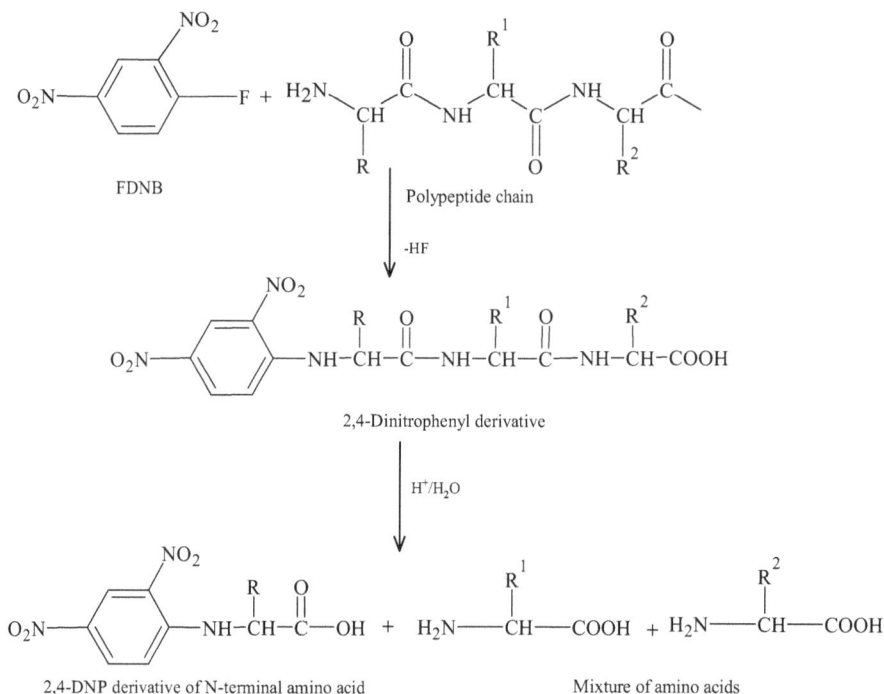

This reaction can also occur with ε-amino group of lysine, hydroxyl group of tyrosine, thiol group of cysteine, and imidazole nucleus of histidine. So, complete hydrolysis of DNP-peptide chain gives DNP derivative of the N-terminal residue along with these amino acid DNP derivatives. 2,4-Dinitrophenyl derivatives of all amino acids have been prepared and characterized, and hence these can be identified either by thin-layer chromatography separation or by determination of the melting points.

ii. The Dansyl method

This method is a modification of DNP method in which 5-(dimethylamino) naphthalene-1-sulfonyl chloride (dansyl chloride, DNS-Cl) is used in place of FDNB. In this method, the polypeptide chain reacts with DNS-Cl at the N-terminal (amino end). Polypeptide chain on hydrolysis results in formation of dansylamino acid and a mixture of free amino acids. This method is widely used because dansyl derivatives give strong fluorescence on irradiation with ultraviolet light. This method is about 100 times more sensitive than the DNP method, and very minute amounts can be easily detected by fluorimetry methods:

N(CH$_3$)$_2$

R$_1$ O   R$_2$ O   R$_3$ O

+ H$_2$N——CH——C—NH—CH——C—NH—CH——C—NH—

Polypeptide chain

O=S=O

Cl

5-(dimethylamino)naphthalene-1-sulfonyl chloride
(Dansyl Chloride)

OH$^-$ HCl

N(CH$_3$)$_2$

O=S=O   R$_1$ O   R$_2$ O   R$_3$ O

NH——CH——C—NH—CH——C-NH—CH——C————

Dansyl polypeptide

H$_2$O

N(CH$_3$)$_2$

O=S=O
       R$_1$ O
NH——CH——C—OH

Dansylamino acid
(Fluorescent)

+   H$_3$N$^+$——CH—COOH + H$_3$N$^+$——CH—COOH
           R$_2$ R$_3$

Free Amino Acids

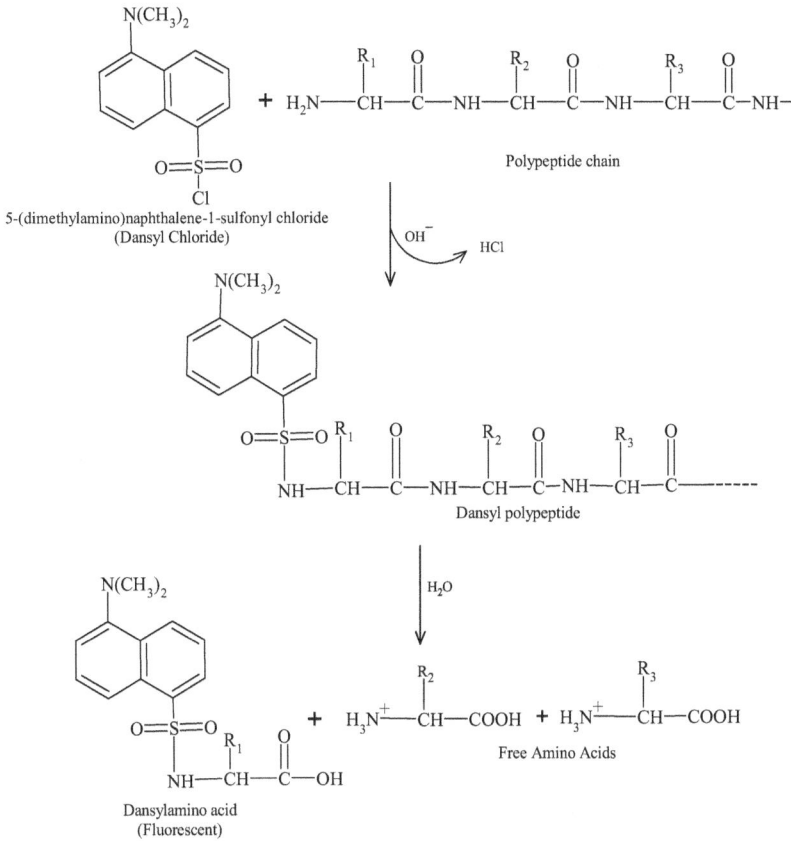

iii. The Edman degradation method

The Edman degradation is one of the most important and widely used methods employed for identifying N-terminal amino acid of a polypeptide chain. This method was introduced by Pehr Edman, which involves the selective removal of N-terminal amino acid leaving rest of the polypeptide chain intact. The process can be repeated for the stepwise degradation of the peptide, and hence, the sequence of amino acid can be elucidated. The Edman method involves the treatment of polypeptide with phenyl isothiocyanate in the presence of dilute alkali to form N-phenylthiocarbamoyl derivative which then treated with HCl in an anhydrous solvent. Further, the N-terminal amino acid is cleaved from the peptide chain and the product, thiazolone, rearranges to 3-phenyl-2-thiohydantoin (PTH) derivative which is then separated and identified by high-performance liquid chromatography. The process is then repeated on the remaining part of the peptide chain to identify the second amino acid and so on. This method has been automated, which has enhanced its efficiency for sequencing the amino acid in polypeptides:

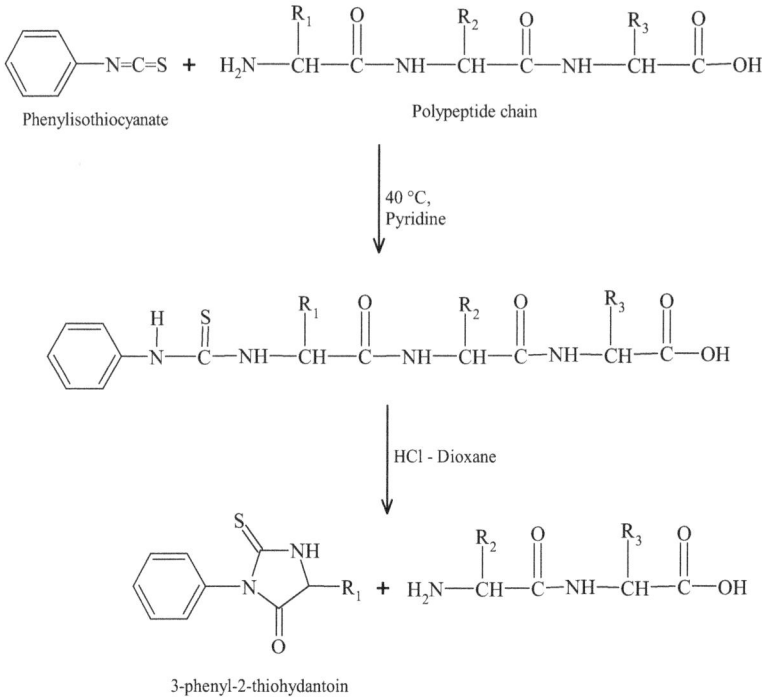

Phenylisothiocyanate

Polypeptide chain

40 °C, Pyridine

HCl - Dioxane

3-phenyl-2-thiohydantoin

2.9.1.2 Carboxyl-Terminal (C-Terminal) Analysis

The identification of C-terminal can be carried out more efficiently by enzymatic methods rather than chemical methods. Although several chemical methods have been investigated, only few of them are extensively used. Some of these methods are:

i. Reaction with benzylamine

In this method, the peptide is converted to an ester and then treated with benzyl-amine to yield benzylamide. The hydrolysis of peptide chain results in the formation of benzylamide of C-terminal amino acid which can be easily characterized:

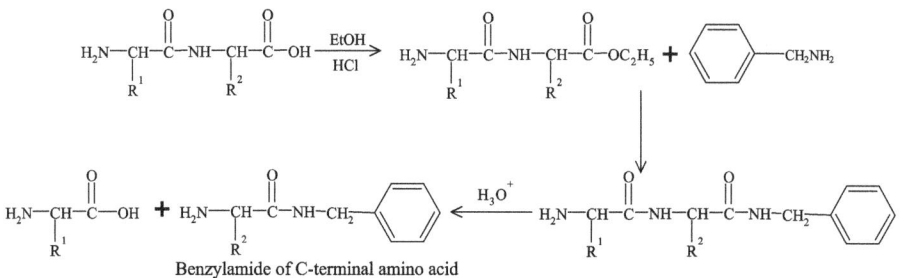

Benzylamide of C-terminal amino acid

ii. Reaction with ammonium isothiocyanate (thiohydantoin formation)

This method of C-terminal identification is similar to the Edman degradation method. On treatment with ammonium isothiocyanate in the presence of acetic anhydride, the C-terminal amino acid is converted into a thiohydantoin. The thiohydantoin can be liberated from rest of the peptide chain using cold alkali. However, the free amino group must be protected first before carrying out this reaction:

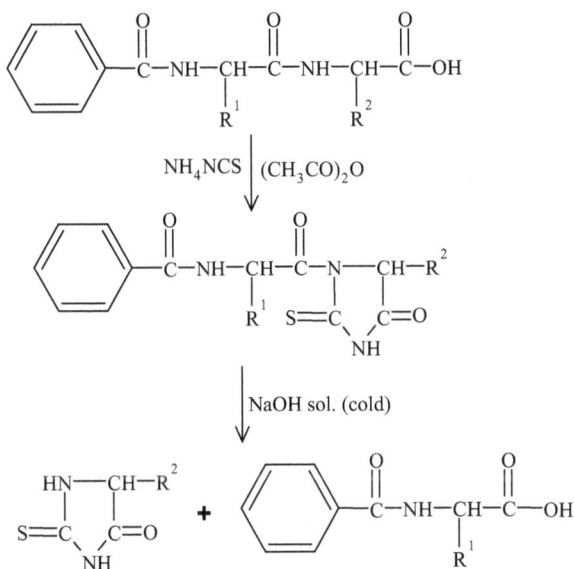

iii. Reaction with hydrazine (hydrazinolysis)

This is one of the most widely used method employed for the identification of C-terminal amino acid, and involves heating of polypeptide chain with anhydrous hydrazine at 100 °C. This cleaves the peptide bonds and forms the hydrazide of all amino acids except the one present at the C-terminal. The mixture of amino acid is then subjected to column chromatography using cation-exchange resin in which the free C-terminal amino acid is eluted and can be identified, whereas the basic hydrazides are retained on the column:

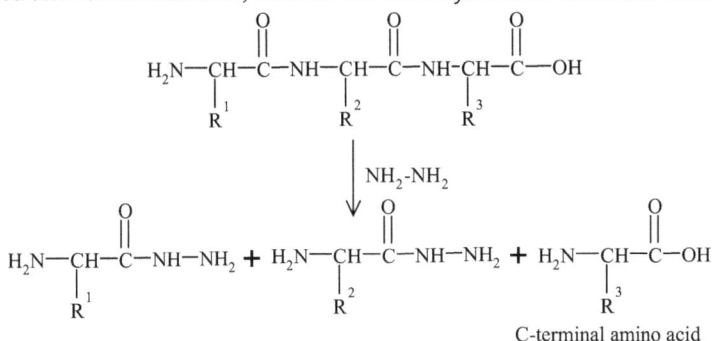

2.9.1.3 Enzymatic Methods

The C-terminal amino acid can also be identified by treating the peptide with enzyme carboxypeptidase, which specifically cleaves the peptide bond from C-terminal only. The removal of C-terminal amino acid yields new shortened polypeptide chain as shown:

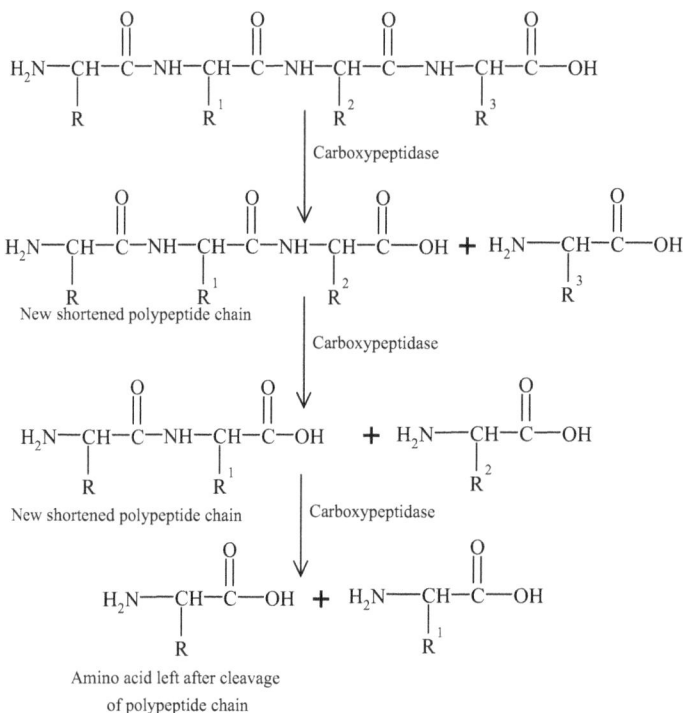

This shortened polypeptide chain with new C-terminal can again be treated with carboxypeptidase. There are two pancreatic enzymes, carboxypeptidase A and carboxypeptidase B, which differ in their reactivity. Carboxypeptidase A cleaves off aromatic amino acids, whereas carboxypeptidase B liberates only basic amino acids, lysine, and arginine.

The stepwise continuous removal of the terminal residues to identify a polypeptide chain is a difficult process. Therefore, the sequence of amino acid can also be determined by carrying out the partial hydrolysis of the polypeptide chain by enzymes. There are specific enzymes which carry out the hydrolysis of the polypeptide chain at specific sites only and form smaller fragments which can then be easily identified:

(i) Trypsin cleaves at the carboxyl end of lysine and arginine.

(ii) Chymotrypsin cleaves at the carboxyl end of phenylalanine, tyrosine, and tryptophan.

(iii) Pepsin is less specific and cleaves at the carboxyl end of leucine, aspartic acid, glutamic acid, phenylalanine, tyrosine, and tryptophan.

(iv) Clostripain carries out selective hydrolysis and cleaves at the carboxyl peptide bond of arginine.

(v) Cyanogen bromide cleaves at the carboxyl end of methionine.

2.10 Proteins

Proteins are biological macromolecules which are essential for the growth and maintenance of living organisms and constitute approximately 75% of the dry material of most living systems. These are synthesized by plants using CO_2, H_2O, NO_3^-, and NH_4^+ in the presence of sunlight (photosynthesis). As discussed earlier, these polymers are made up of large number of amino acids joined together by peptide bonds, and have three-dimensional structures. The function of proteins can be well understood by considering its structure at four different levels, namely, primary, secondary, tertiary, and quaternary. The arrangement of the polypeptide in a protein molecule is determined by its primary structure (the amino acid sequence), whereas the conformation that the polypeptide chain assumes is called the secondary structure. The way the molecule folds to produce a specific shape is called a tertiary structure of the protein, and quaternary structure describes the arrangement and ways in which the subunits of proteins are held together.

2.10.1 Primary Structure

The primary structure of a protein is the linear arrangement of amino acids, linked by peptide bonds to form a polypeptide chain (Figure 2.8). It refers to the number, nature, and sequence of amino acids present in its polypeptide chain. The various chemical/biological properties of the protein are determined by its primary structure.

2.10.2 Secondary Structure

The secondary structure of the protein is the spatial arrangement of polypeptide chains. The amide bond is a rigid bond but free rotation between C–C single bond and C–N single bond in amino acid side chain leads to different conformations of the polypeptide chain resulting in the formation of secondary structure of the protein. In this structure, hydrogen bonding occurs between carboxyl oxygen and hydrogen atom of amino group of the peptide chain. These bonds can occur either within the same molecule of the polypeptide chain or between different polypeptide chains. The two most common secondary structures adopted by polypeptide chains are alpha-helix (α-helix) and beta-pleated sheets (β-pleated sheet), though beta (β) turn and omega (Ω) loop also occur.

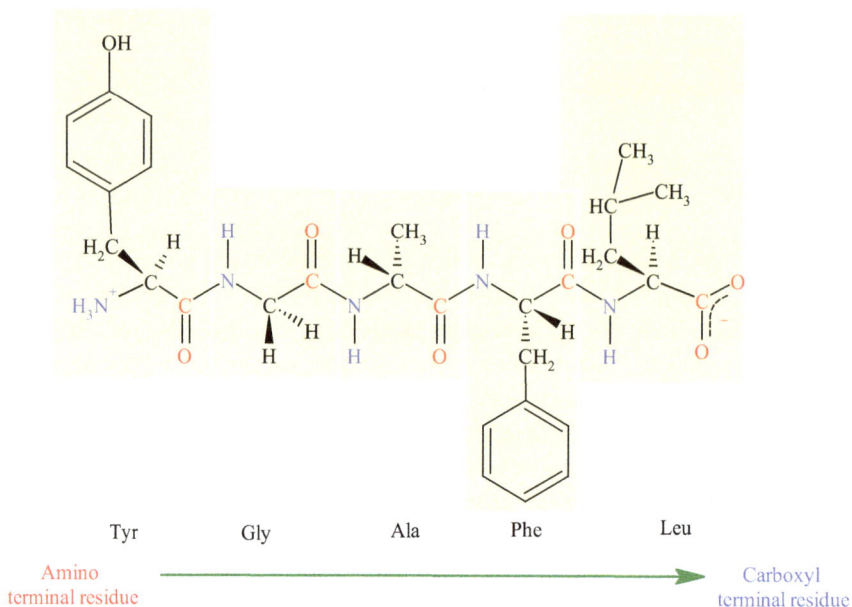

Figure 2.8: Primary structure of a protein.

i. α-Helix: In 1951, Linus Pauling and Robert Corey proposed two structures known as the α-helix and the β-pleated sheet. The α-helix, the first proposed structure, is a rodlike structure in which tightly coiled backbone forms the inner part of the rod, whereas side chain protrudes in a helical fashion. The helical structure is stabilized by hydrogen bonding between the –C = O group of the first amino acid and –NH– of the fourth amino acid (Figure 2.9). Thus, all the main chain –C = O and –NH– are hydrogen bonded except the amino acid present near the end of the helix.

In an α-helix, the polypeptide backbone follows a helical path and the distance between adjacent amino acids is 1.5 Å (also called translation) along the helical axis. Each turn of the helix has 3.6 amino acid residues, and the pitch of the helix, that is, the length of one complete helical turn, is 5.4 Å. It is equal to the product of translation (1.5 Å) and number of amino acid residues per turn (3.6). The α-helix can be right-handed (clockwise) or left-handed (anticlockwise) but the helices of most of the natural proteins are right-handed. In 1956, Moffitt deduced that the right-handed helix (for L-amino acid) is more stable than the left-handed helix which is also supported by the fact that except glycine, all amino acids are optically active and have the L-configuration. In an α-helix, the functional groups containing side chain R-groups (side chain groups) protrude out from the helically coiled polypeptide backbone. R-groups of amino acid residues are exclusively present on the surface of a helix. Among all amino acids, only few have a great propensity to exist in an α-helix; however, all cannot be easily accommodated in helical form. Valine, threonine, and isoleucine have branching at the β-carbon atom

(a)

(b)

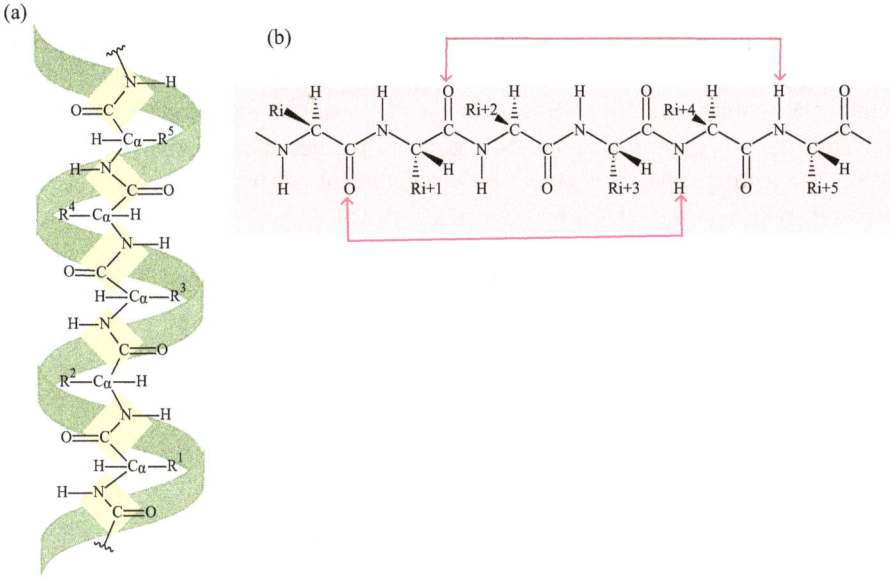

Figure 2.9: (a) Structure of the α-helix and (b) hydrogen bonding scheme in an α-helix.

which destabilize α-helices because of steric hindrance. The presence of serine, aspartate, and asparagine is also prone to disrupt α-helices because these amino acids contain hydrogen bond donors and acceptors in their side chains which compete with the main chain –C = O and –NH– groups for hydrogen bonding. Proline also tends to interrupt an α-helix because it lacks –NH– group, and its ring structure prevents the polypeptide backbone from adopting a conformation compatible with an α-helix. The α-helical structure is present in many proteins, for example, it is found in ferritin, a protein that stores iron. It is also the predominant structure of the polypeptide chain of myosin, a muscle protein, and α-keratin, a fibrous structural protein, present in hair, horns, nails, and mammalian claws.

ii. β-Pleated sheet (β-conformation): This is the second type of structure which was proposed by Corey and Pauling in 1951 after α-helix, hence, named β. In this secondary structure, two or more polypeptide chains called strands are connected to each other through intermolecular hydrogen bonding. β-sheet consists of strands of protein lying adjacent to one another, interacting laterally via hydrogen bonds between backbone –C = O and –NH– of amino acids. The distance between adjacent amino acids in a β-strand is approximately 3.5 Å. The adjacent polypeptide chains are either parallel or antiparallel to each other. In a parallel β-pleated sheet, all the polypeptide chains run in the same direction, that is, N-termini of both strands at the same end, whereas in antiparallel β-pleated sheet, the chains run in opposite direction. In the parallel arrangement, hydrogen bonding between two adjacent strands is little complicated as the –NH– group on one strand is hydrogen bonded to the –C = O group of one amino acid (say first) on the adjacent strand, whereas the –C = O group is hydrogen bonded to the –NH– group on the third amino acid along the chain (two residues farther). In the antiparallel arrangement, the –NH– and the –C = O groups on one strand are hydrogen bonded to the –C = O group and the –NH– group of partner amino acid on the adjacent strand. The hydrogen bonding pattern between adjacent polypeptide chains in parallel and antiparallel β-pleated sheets is shown in Figure 2.10.

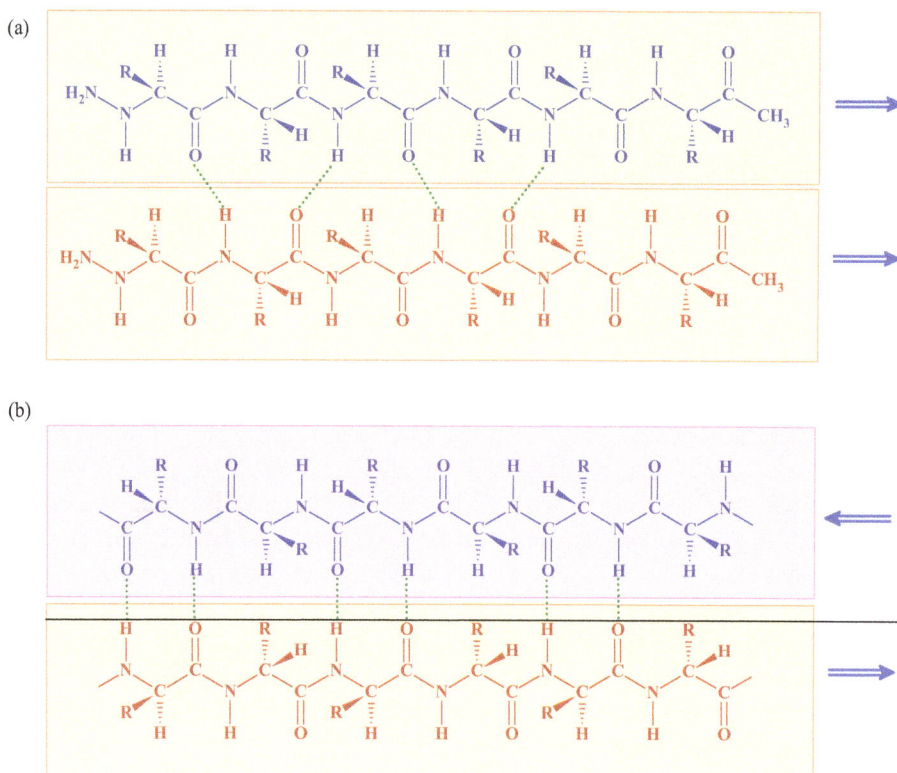

Figure 2.10: Structure of (a) a parallel β-pleated sheet and (b) an antiparallel β-pleated sheet.

Along with the parallel and antiparallel arrangement, these structures can also adopt mixed conformation in which two strands run parallel to each other, whereas antiparallel to the third strand or vice versa (Figure 2.11). β-Sheet is puckered due to the tetrahedral nature of carbon bonds, hence, named pleated sheet. R groups of amino acids in a β-strand alternatively point to one side or the other of a β-strand. This is the reason every other amino acid is exposed on one side or the other of a β-sheet. Alanine, valine, methionine, and isoleucine readily form β-pleated sheets. β-Keratin, protein that is found in feathers and claws, is an example of this secondary structure. Silk, a fibroin protein, also contains large segments of β-pleated sheets.

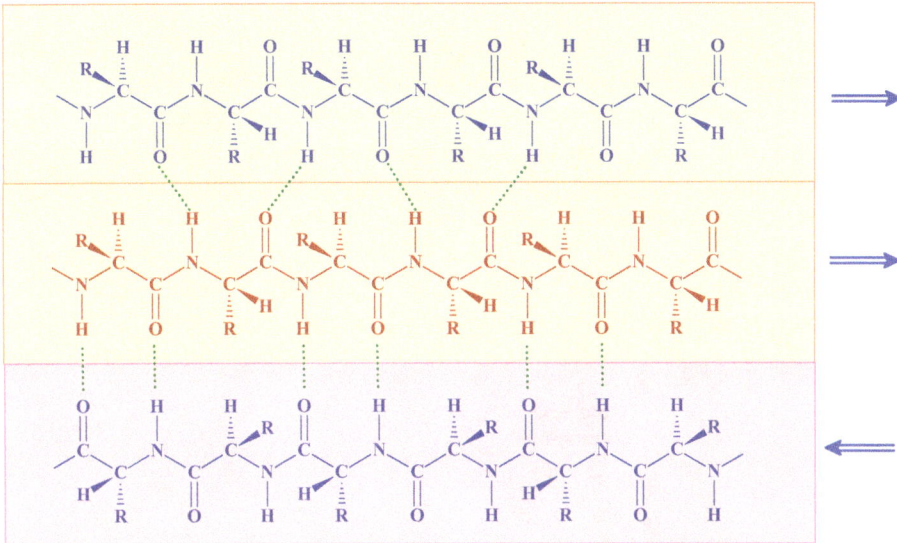

Figure 2.11: Structure of a mixed β-sheet.

2.10.3 Tertiary Structure

Tertiary structure of the protein refers to the three-dimensional structure that describes the overall spatial arrangement of the polypeptide chain. The physiological three-dimensional conformation of a protein is maintained by a range of noncovalent interactions (electrostatic forces, hydrogen bonds, and hydrophobic forces) and covalent interactions (disulfide bonds) in addition to the peptide bonds between individual amino acids. So, a polypeptide chain tends to fold in a fashion that maximizes the number of stabilizing forces, and these are as follows:

(i) **Electrostatic interactions:** These include the interactions between two oppositely charged ions or groups, for example, the ammonium group of lysine (Lys) and the carboxyl group of aspartate (Asp), often referred to as an ion pair or salt bridge.

(ii) **Hydrogen bonding:** In biological molecules, the donor is a hydrogen atom, and the acceptor is either oxygen or nitrogen. Hydrogen bonds are stronger than van der Waals forces but they are much weaker than covalent bonds. Hydrogen bonding not only plays an important role in stabilizing protein structure, but it is also one of the major factors that stabilize DNA double helix and lipid bilayers. Serine and threonine possess hydroxyl groups so these polar amino acids can form hydrogen bonds. Asparagine and glutamine having amide bonds in their side chains are usually in hydrogen-bonded form, stabilizing the tertiary structure.

(iii) **Hydrophobic forces:** Hydrophobic interactions also play an important role in folding of a polypeptide chain. Hydrophobic side chains of nonpolar amino acids can interact with each other through weak van der Waals interaction.

(iv) **Disulfide bonds:** These covalent bonds form between two cysteine (Cys) residues that are close to each other in the final conformation and form cystine that provides stability to its three-dimensional structure. These bonds can be formed within a single polypeptide chain or between different polypeptide chains.

The tertiary structure of a protein is shown in Figure 2.12. Myoglobin, albumin, plasma proteins, and the immunoglobulins are some of the proteins which possess the tertiary structure.

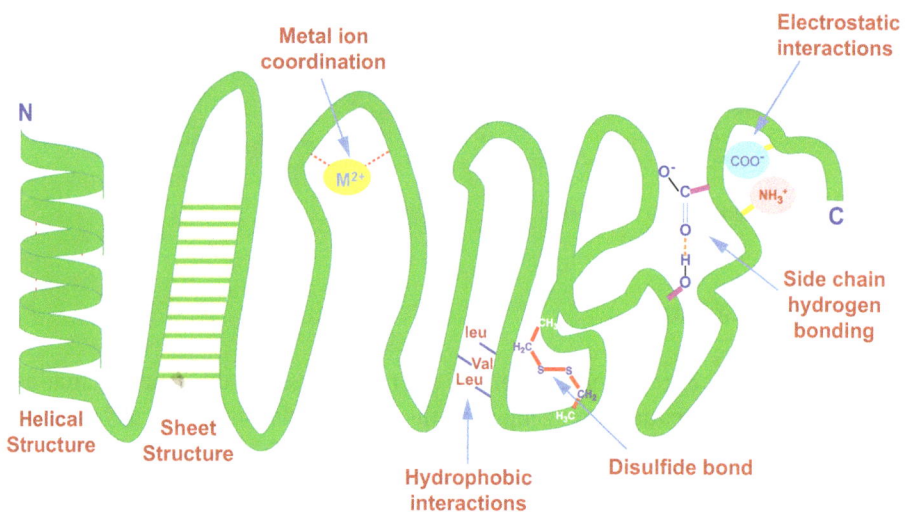

Figure 2.12: Tertiary structure of a protein.

2.10.4 Quaternary Structure

The quaternary structure of the protein is formed with the noncovalent association between two or more polypeptide chains as shown in Figure 2.13. Two or more identical polypeptide chains or different polypeptide chains are involved in the formation of multisubunit protein. Both the fibrous and globular proteins consist of only one polypeptide chain. If several chains are present, the globular protein is said to be oligomeric. The individual chains are known as protomers or subunits, which may or may not be identical. These subunits are held together by same interactions that hold the individual polypeptide chain in the tertiary structure. Myoglobin, which possesses a tertiary structure, consists of a single polypeptide chain, which contains about eight straight subunits (α-helices), and is folded in an irregular manner at the random coil section. On the

Figure 2.13: Quaternary structure of a protein.

other hand, hemoglobin contains four subunits, two identical α-chains and two β-chains. Each of those subunits has a tertiary structure similar to that of myoglobin.

2.11 Classification of Proteins

Proteins are one of the most important versatile biological molecules which perform essential functions in living organisms. They exhibit a wide range of structural and functional diversity that it is not possible to classify them in one single category. They can be classified based on their shape and solubility, structure, and functions.

2.11.1 Classification of Proteins Based on Their Shape

Based on their shape and solubility, proteins are classified into two major categories:

(a) Fibrous proteins – Fibrous proteins consist of large helical content, and are rigid molecules with rod-like shape. They serve as a main structural component of animal tissues. Fibrous proteins are insoluble in water. Example of fibrous proteins are collagen, keratin, myosin, elastins, and fibroin.

(b) Globular proteins – Globular proteins are folded into spherical shape, and due to the distribution of amino acids (hydrophobic inside, hydrophilic outside), they are soluble in aqueous solution (form colloids in water), unlike the fibrous proteins. Globular proteins perform functions that require mobility and hence solubility. Also, they are involved in maintenance and regulation of life processes. Examples under this category include hemoglobin, insulin, myoglobin, and fibrinogen.

2.11.2 Classification of Proteins Based on Their Structure

Proteins can be classified into three categories based on their structure:

(a) **Simple proteins:** Simple proteins are composed of only amino acid residues and do not contain any nonprotein part. They yield only amino acids on hydrolysis and the major type of simple proteins are albumins (ovalbumin, serum albumin, casein in milk, and gliadin (wheat protein)), globulins, protamines, histones, and glutelins.

(b) **Conjugated proteins:** These proteins consist of simple globular protein and some nonprotein part. The nonprotein group is called prosthetic group and is very important for the functioning of protein. Depending on the nature of prosthetic group, conjugated protein can be further classified as follows:
(i)   Nucleoproteins: In nucleoproteins, the protein molecules are combined with the nucleic acids, for example, nucleohistones and nucleoprotamines.
(ii) Glycoproteins (mucoproteins): Glycoproteins consist of simple protein combined with carbohydrates. Examples in this category include mucin (saliva), heparin (bile juices), and mucopolysaccharides of cartilage and tendons.
(iii) Lipoproteins: Proteins found in combination with lipids are termed as lipoproteins. These proteins are present in the brain, milk, and egg yolk.
(iv) Phosphoproteins: These proteins contain phosphoric acid as the prosthetic group, and casein (milk protein) and ovovitellin are the examples of phosphoproteins.
(v)   Metalloproteins: These proteins contain some metal ions like iron, copper, zinc, and cobalt. Examples in this category include carbonic anhydrase (Zn) and ceruloplasmin (Cu).
(vi) Chromoproteins: In chromoproteins, simple protein consists of colored pigment as prosthetic group. Hemoglobin, myoglobin, and cytochromes are some examples in this category.

(c) **Derived proteins:** These are the degraded or denatured products obtained on the acidic/alkaline/enzymatic hydrolysis of simple and conjugated proteins. These are also considered as intermediate hydrolysis products. These proteins are further divided into two classes: primary derived proteins and secondary derived proteins.

(i) Primary derived proteins
1. **Coagulated proteins:** These are the coagulated or denatured proteins produced by the action of heat, acids, or alkalis on proteins. In these proteins, molecular mass does not change, and only physical properties such as solubility and precipitation change; examples are cooked proteins and coagulated albumin (egg white).
2. **Proteans:** Proteans are the earliest products of protein hydrolysis formed by the action of water, dilute acids, and enzymes. These are generally obtained from globulins and are insoluble in water and dilute salt solutions, for example, fibrin from fibrinogen.

3. **Metaproteins:** These are the second-stage hydrolysis products of proteins and require slightly stronger acids and alkalis. These are found to be insoluble in water or dilute salt solution but soluble in acids as well as in alkalis. These can be precipitated by half saturation with ammonium sulfate.

(ii) Secondary derived proteins

These proteins are the progressive hydrolytic products of the peptide bonds of protein molecule. They include proteoses, peptones, and peptides according to the average molecular weight.

2.11.3 Classification of Proteins Based on Their Functions

Proteins can be classified into the following categories based on their functions:

(i) **Enzymes or catalytic proteins**: Enzymes are biocatalysts which catalyze specific biochemical reactions in the cell and most of the enzymes are proteins. Examples are pepsin, chymotrypsin, and trypsin.

(ii) **Structural proteins:** These proteins impart strength to biological structures or protect organisms from their environment. For example, collagen is the most abundant protein found in animals which form a major constituent of skin, bones, muscles, cartilage, and tendons. Keratin is the major component of hair, hooves, feathers, fur, claws, and the outer layer of skin.

(iii) **Contractile proteins:** Proteins like actin and myosin are involved in the process of muscle contraction.

(iv) **Protective (defense) proteins:** Snake venoms and plant toxins protect their owners from predators. Blood-clotting proteins protect the vascular system when it is injured.

(v) **Hormones**: Some of the hormones, such as insulin, that regulate the metabolic processes is a protein.

(vi) **Transport and storage proteins:** These proteins are involved in the process of transport and storage of biological substances. The best example is oxygen-carrying protein – hemoglobin. Some proteins act as storage proteins, for example, ferritin stores iron in the liver.

2.12 Denaturation of Protein

We have seen that proteins play a crucial role in various biological processes and are important for the growth and development of an individual. This versatile molecule must remain in biologically active form at all levels for proper functioning. Protein molecules may undergo a change in their structure if they are subjected to heat or alteration in their environmental conditions. This process of the disruption of bonds and change in overall structure of a protein is termed as denaturation. It results in the loss of secondary, tertiary, and quaternary structures of proteins and alters the physical, chemical, and biological properties of the protein molecule (Figure 2.14). If proteins in a living cell are denatured, this disrupts the cell activity and may lead to cell death.

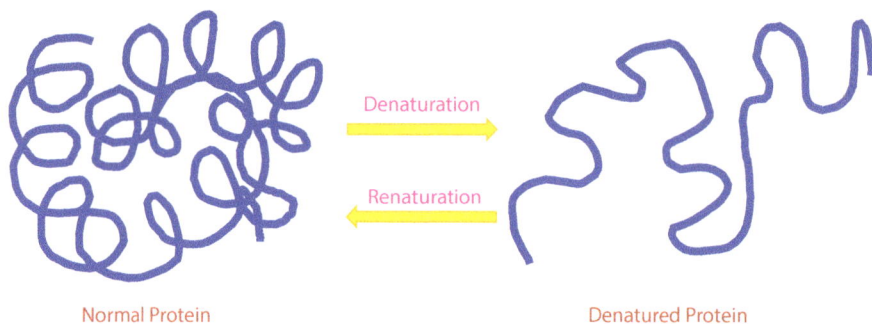

Figure 2.14: Denaturation of protein.

Denaturation conditions are not strong enough to break the peptide bonds, so the primary structure of protein remains intact even after denaturation process. It disrupts the secondary structures and eventually turns α-helix and β-sheet into a random coil. In tertiary structure, four types of interactions are present between side chains, including hydrogen bonding, ionic interactions, disulfide linkages, and nonpolar hydrophobic interaction. All these interactions can be disrupted on denaturation, and protein molecules lose their biological activity. Denaturation can be carried out by various physical (heat, X-rays, UV light, and high pressure) and chemical (acids, alkalis, alcohol, heavy metal salts, urea, and guanidinium chloride) methods. Alcohol disrupts hydrogen bonds between side chains, and new hydrogen bonds are formed between alcohol and the side chains of proteins as shown in Figure 2.15.

Figure 2.15: Disruption of hydrogen bonding by alcohol.

Ionic interactions (salt bridges) are formed between positively charged ammonium group and the negative acid group, and acids and bases might disrupt these salt bridges held together by ionic interactions (Figure 2.16).

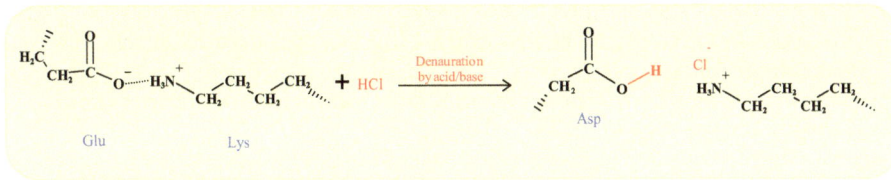

Figure 2.16: Disruption of ionic interaction by acid.

Heavy metal also disrupts disulfide bonds in protein as they have high affinity for sulfur which results in the denaturation of proteins. The disulfide bonds can also be cleaved by β-mercaptoethanol, which converts disulfides (cystines) into sulfhydrals (cysteines) (Figure 2.17). Urea and guanidinium chloride also disrupt the protein non-covalent bond and unfold the protein.

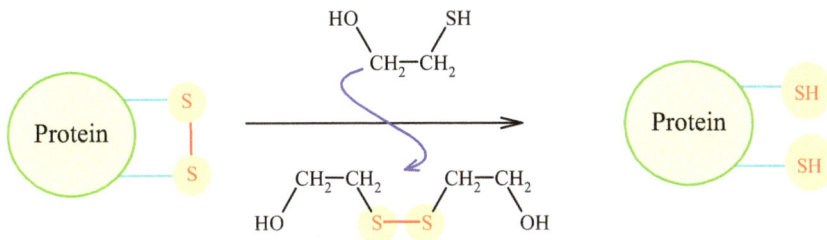

Figure 2.17: Disruption of disulfide linkages by β-mercaptoethanol.

References

[1] Ahluwalia VK, Goyal M. A Textbook of Organic Chemistry. Narosa Publishing House; India 2000.
[2] Berg JM, Tymoczko JL, Gatto GJ. Jr Lubert Stryer. Biochemistry. Freeman Macmillan. 8th Edition. ISBN-13. 2006:978–1.
[3] Finar IL. Organic Chemistry (Volume 2), Dorling Kindersley (India) Pvt. Ltd 1964.
[4] Isidro-Llobet A, Alvarez M, Albericio F. Amino acid-protecting groups. Chemical Reviews. 2009 Jun 10;109(6):2455–504.
[5] Mehta B, Mehta M. Organic Chemistry. PHI Learning Pvt. Ltd.; India 2015 Aug 31.
[6] Murray K, Rodwell V, Bender D, Botham KM, Weil PA, Kennelly PJ. McGraw Hill Harper's Illustrated Biochemistry. 28. Citeseer, New York, United States. 2009.
[7] Nelson DL, Lehninger AL, Cox MM. Lehninger Principles of Biochemistry. Macmillan; 2008. Lehninger AL, Nelson DL, Cox MM. Lehninger Principles of Biochemistry. Fourth ed. New York: W.H. Freeman; 2005.
[8] Press CR. Handbook of Biochemistry and Molecular Biology. CRC Press; Boca Raton, Florida. 2010 May 21.

[9]   Snyder SA, Fryhle CB, Solomons TW. Organic Chemistry. Graham Solomons T.W., Fryhle C.B., Snyder S.A. Solomons'. Amino Acids and Proteins. Chapter 24. pp 1046–1085. Organic Chemistry. 2016; 12th edn, global edition Wiley

[10] Sureshbabu VV, Narendra N. Protection reactions. Amino acids, peptides, and proteins in organic chemistry. 2011 Mar 18;4:1–97.

Chapter 3
Enzymes

The chemical reaction, that is, the transformation of one molecule to different molecule, that takes place inside the cells of a living organism is known as biochemical reaction, and the chemistry involved in these conversions is quite complex. Metabolism is a series of biochemical reactions that allow the cell to function properly, and helps an individual to grow, reproduce, and stay alive. Our cells are composed of thousands of chemicals, which are made up of elements such as carbon, hydrogen, oxygen, nitrogen, phosphorus, and sulfur. The conversion of one chemical into other takes place via metabolic pathways, and each step in these pathways is facilitated by a specific enzyme. Enzymes are catalytic proteins that catalyze the biochemical reactions which take place within the cells (hence the name biocatalysts). Almost all biochemical reactions would occur extremely slow if suitable enzymes are not present in the biological systems to speed them up, and life would be impossible without them. A biochemical reaction could take up several days or weeks to occur without an enzyme but the same reaction may get completed in few seconds with the help of enzymes. They can catalyze several million reactions within a second. In the late eighteenth and early nineteenth centuries, the processes like digestion of meat by stomach secretions and the conversion of starch into sugars were known but the mechanisms behind these processes had not been understood. In the nineteenth century, Louis Pasteur, while carrying out the fermentation of sugar to alcohol by yeast, stated that the fermentation process is catalyzed by some force present within the yeast cells and function only within living organisms. Although the first enzyme, diastase, was discovered by Anselme Payen in 1833, the term "enzyme" (meaning "in leaven") was coined by the German physiologist Wilhelm Friedrich Kühne in 1878. In 1897, Eduard Buchner showed that living cells were really not required for the fermentation of sugar into alcohol and carbon dioxide. He showed that an enzyme zymase can be extracted from yeast cells to carry out the fermentation of sucrose. In 1907, Buchner received Nobel Prize in Chemistry for this discovery. In 1926, James B. Sumner crystallized the enzyme "urease," and 4 years later, John H. Northrop and Wendell M. Stanley isolated pepsin, trypsin, and chymotrypsin, and revealed that enzymes are proteins. These three scientists were awarded the Nobel Prize in Chemistry in 1946 for their pioneer work in the field of enzymes.

Enzymes are involved in almost all biochemical reactions, and they perform their job with great accuracy and at astonishing speed. Without enzymes, most of the biochemical reactions do not occur at the perceptible rates, and even a simple reaction as hydration of carbon dioxide requires an enzyme. Carbonic anhydrase is one of the fastest enzymes known, and can hydrate 10^6 molecules of CO_2 per second. One of the most important functions of enzymes is to help in the digestion of food. Digestive enzymes break down the large complex molecules such as carbohydrates, proteins, and fats into smaller

https://doi.org/10.1515/9783110793765-003

ones, and allow the nutrients obtained from food to be absorbed by the body. Without these digestive enzymes, animals would not be able to break down these complex molecules and would not get energy and nutrients which they require for their survival. In enzymatic reactions, the substrate molecules are converted into different molecules, known as products. These biocatalysts increase the rate of reaction without themselves undergoing any change. Their catalytic power is extremely high in comparison to that of inorganic catalyst. Almost all chemical reactions in a biological cell needs enzyme in order to occur at rates which are crucial for the existence of life, and they are known to catalyze more than 5,000 biochemical reactions. Nearly all enzymes are proteins, except few, which are catalytic RNA molecules (called ribozymes).

3.1 Classification of Enzymes

There are different ways of naming and classifying an enzyme. Mostly, the enzymes have been named by adding suffix *–ase* to the name of the substrate, while some are named by a word describing their activity. The first active enzymes, *zymases*, were found in yeast. *Zymase* catalyzes the fermentation of glucose to ethyl alcohol and carbon dioxide. Urease catalyzes the hydrolysis of urea, and DNA polymerase carries out the synthesis of DNA in the cell. The names of some of the enzymes also describe their functions like *oxidases* catalyze oxidation reactions, while digestive enzymes such as *pepsin* and *trypsin* do not carry the substrate name or suffix *–ase* and help in digestion. Sometimes, two different enzymes can have the same name or one enzyme can have more than two names. To remove such ambiguity and for uniformity in naming, the International Commission on Enzymes was established by the International Union of Biochemistry (now known as the International Union of Biochemistry and Molecular Biology, IUBMB), which devised a common system and nomenclature for classification of enzymes in 1961. According to IUBMB, enzymes have been classified into six classes on the basis of the reaction they catalyze, and each class is identified by an EC (Enzyme Commission) number. EC numbers are represented by four components (say 1.2.3.4) separated by periods, where first number (1) shows the main class to which the enzyme belongs, the second figure (2) is the subclass, the third figure (3) gives the sub-subclass, and the fourth figure (4) is the sub-sub-subclass. The subclass and sub-subclass describe the reaction, while the sub-sub-subclass is a serial number used to identify a particular enzyme within a sub-subclass (based on the actual substrate in the reaction). This classification does not consider the homology of amino acid sequences, structure of protein, or chemical mechanism involved. For example, *alcohol dehydrogenase* (EC number is 1.1.1.1), belonging to the oxidoreductase family, catalyzes the oxidation of ethanol to acetaldehyde using NAD^+ (nicotinamide adenine dinucleotide) and $NADP^+$ (nicotinamide adenine dinucleotide phosphate). The systematic name of this enzyme is alcohol:NAD^+ oxidoreductase. However, it was noticed that none of these classes define the enzymes that carry out the movement of ions across the cell membranes, and hence, a new class of enzyme, translocases (EC7), was added in 2018. The classification of enzymes and their examples are given in Table 3.1.

Table 3.1: Classification of enzymes.

Enzyme class	Type of reaction catalyzed	Examples
EC1 Oxidoreductases	Catalyze redox reaction, categorized into oxidases and reductases	Dehydrogenases, oxidases
EC2 Transferases	Catalyze transfer of a specific functional group from one molecule to another molecule	Transaminases, kinases
EC3 Hydrolases	Catalyze hydrolysis of a substrate	Esterases, glycosylases
EC4 Lyases	Catalyze the removal of a group from a substrate to form a double bond or the reverse reaction	Decarboxylases, aldolases
EC5 Isomerases	Catalyze the geometrical or structural changes within one molecule (isomerization reactions)	Racemases, epimerases, isomerases
EC6 Ligases	Catalyze the joining of two molecules coupled with the hydrolysis of ATP	DNA ligases, aminoacyl-tRNA synthetase
EC7 Translocases	Catalyze the movement of ions or molecules across cell membranes or their separation within membranes	Proton-translocating $NAD(P)^+$ transhydrogenase, ABC-type phosphate transporter

EC1 Oxidoreductases: This class of enzymes catalyzes the transfer of electron from one molecule to another molecule, usually taking NAD^+ as cofactor. *Oxidases, dehydrogenases, reductases, oxygenases, peroxidases*, and *hydroxylases* belong to this class of enzymes. The systematic names of oxidoreductases are in the form of "donor:acceptor oxidoreductase"; however, "donor:dehydrogenase" is commonly used. Oxidases are used in cases where O_2 acts as an acceptor:

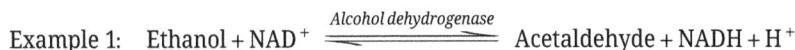

Example 1: $\text{Ethanol} + \text{NAD}^+ \xrightleftharpoons{\textit{Alcohol dehydrogenase}} \text{Acetaldehyde} + \text{NADH} + \text{H}^+$

Example 2: $\text{Pyruvate} + \text{NADH} + \text{H}^+ \xrightarrow{\textit{Lactate dehydrogenase}} \text{Lactate} + \text{NAD}^+$

EC2 Transferases: These enzymes catalyze the transfer of specific functional groups from one molecule (generally regarded as donor) to another (generally regarded as acceptor). These are named according to the form "donor:acceptor group transferase"; nonetheless, more frequently used nomenclature for this class is "acceptor group transferase" or "donor group transferase." For example, the systematic name of hexo-kinase enzyme that catalyzes the conversion of glucose to glucose-6-phosphate is ATP: D-hexose 6-phosphotransferases. *Transaminases, transaldolases, phosphotransferases,* and so on belong to this category of enzymes:

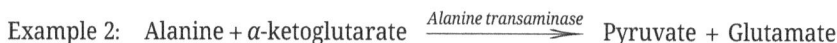

Example 1: $\text{Glucose} + \text{ATP} \xrightarrow{\text{\textit{Hexokinase}}} \text{Glucose} - 6 - \text{phosphate} + \text{ADP} + \text{H}^+$

Example 2: $\text{Alanine} + \alpha\text{-ketoglutarate} \xrightarrow{\text{\textit{Alanine transaminase}}} \text{Pyruvate} + \text{Glutamate}$

EC3 Hydrolases: Hydrolases belong to hydrolytic enzymes which catalyze the hydroly-sis reaction utilizing water to break a bigger molecule into smaller ones. This catalyzes the hydrolytic cleavage of C–O, C–N, and C–C bonds, along with some phosphoric anhy-dride bonds. The systematic name for this class of enzymes is "substrate hydrolase," whereas commonly they are named as "substratease." This category includes *esterases, lipases, phosphatases, peptidases/proteases,* and *nucleosidases.* Depending on the type of bond they act upon, hydrolases can further be categorized into 13 subclasses. These en-zymes are important for the body as they break down the biopolymers (e.g., lipases break down fats) into monomers (fatty acids and glycerol) for synthesis and provide carbon sources for energy production:

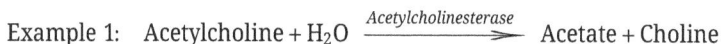

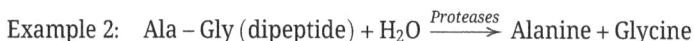

Example 1: $\text{Acetylcholine} + \text{H}_2\text{O} \xrightarrow{\text{\textit{Acetylcholinesterase}}} \text{Acetate} + \text{Choline}$

Example 2: $\text{Ala} - \text{Gly (dipeptide)} + \text{H}_2\text{O} \xrightarrow{\text{\textit{Proteases}}} \text{Alanine} + \text{Glycine}$

EC4 Lyases: These enzymes catalyze the reaction involving the addition of functional groups across a double bond or the reverse reaction, that is, nonhydrolytic removal of functional groups from substrates by elimination, creating a double bond in the prod-uct. Lyases can further be classified into seven subclasses and their systematic name is "substrate group lyase" such as *decarboxylases and aldolases* in the removal direc-tion, and synthases in the addition direction:

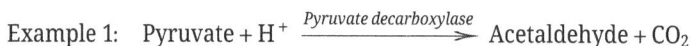

Example 1: $\text{Pyruvate} + \text{H}^+ \xrightarrow{\text{\textit{Pyruvate decarboxylase}}} \text{Acetaldehyde} + \text{CO}_2$

Example 2: $\textit{cis}\text{-Aconitate} + \text{H}_2\text{O} \xrightarrow{\text{\textit{Aconitase}}} \text{Isocitrate}$

EC5 Isomerases: Isomerases are a group of enzymes which catalyze geometrical or structural changes within the molecule, generating isomeric forms. *Isomerases* facili-tate intramolecular rearrangements in which bonds are broken and formed. They have assigned EC 5 and are further categorized into six subclasses. *Isomerases* cata-lyze many important biological reactions such as glycolysis and glucose metabolism.

Racemerases, epimerases, cis–trans isomerases, intramolecular oxidoreductases, tauto-merases, and *mutases* are included in this category:

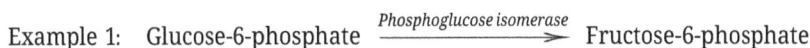

Example 1:   Glucose-6-phosphate $\xrightarrow{\text{Phosphoglucose isomerase}}$ Fructose-6-phosphate

Example 2:   Dihydroxyacetone phosphate $\xrightarrow{\text{Triose phosphate isomerase}}$ Glyceraldehyde $-3-$ phosphate

EC6 Ligases: Ligases, also known as synthetases, are used to catalyze the joining of two molecules by the formation of new chemical bonds (mostly C–C, C–O, C–N, and C–S). These reactions are coupled with the breakdown of energy-containing sub-strates, usually ATP. The common name of these enzymes often has the word *"ligase."* For example, *DNA ligase* catalyzes the joining of two DNA fragments by forming a phosphodiester bond:

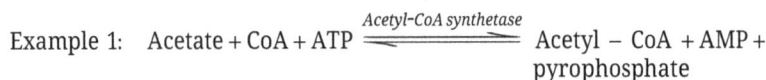

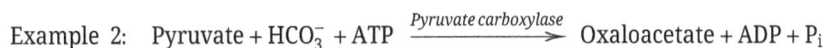

Example 1:   Acetate $+ \text{CoA} + \text{ATP} \xleftrightharpoons{\text{Acetyl-CoA synthetase}}$ Acetyl $-$ CoA $+ \text{AMP} +$ pyrophosphate

Example 2:   Pyruvate $+ \text{HCO}_3^- + \text{ATP} \xrightarrow{\text{Pyruvate carboxylase}}$ Oxaloacetate $+ \text{ADP} + \text{P}_i$

EC7 Translocases: This class of enzyme was added to the list of enzyme classification and nomenclature in 2018. Translocases catalyze the movement of hydrogen ions, in-organic cations and anions, carbohydrates, amino acids, and other molecules across the cell membrane. They also keep them separated within the membrane. These are further classified into six categories describing the types of ions and molecules trans-located. The reaction is shown as a transfer from "side 1" to "side 2" to remove the ambiguity as "in" and "out" had been used earlier:

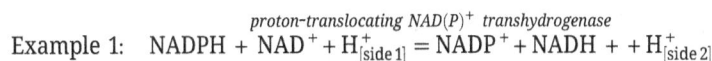

Example 1: $\text{NADPH} + \text{NAD}^+ + \text{H}^+_{[\text{side 1}]} \xrightleftharpoons{\text{proton-translocating NAD(P)}^+ \text{ transhydrogenase}} \text{NADP}^+ + \text{NADH} + + \text{H}^+_{[\text{side 2}]}$

3.2 Specificity of Enzyme Action

One of the main significant properties of an enzyme is its specificity toward one reac-tion over other, which makes them an important tool in research and medical diagnosis. Enzymes are highly specific to the reactions they catalyze and to the substrates involved in these reactions. Their specificity is due to the specific arrangement of atoms in the active site. The weak noncovalent interactions like van der Waals forces, electrostatic forces, hydrogen bonding, and hydrophilic/hydrophobic interactions between enzyme and substrate are responsible for this specificity, that is, it depends on the molecular complementarity. There are four different types of specificity exhibited by enzymes:

i. **Absolute specificity:** In absolute specificity, enzyme catalyzes only one type of reaction for a single substrate. For example,
    (a) *Urease*, a metalloenzyme, catalyzes the hydrolysis of urea into carbon dioxide and ammonia.
    (b) *Sucrase* is a digestive enzyme that catalyzes the hydrolysis of sucrose into fructose and glucose.
ii. **Group specificity:** This type of specificity is exhibited by enzymes which are specific to a specific type of bond and groups surrounding the bonds. Example includes:
    (a) *Trypsin* is an endopeptidase that cleaves peptide chains mainly at the carboxyl side of the amino acids, lysine and arginine.
    (b) *Chymotrypsin*, another endopeptidase, catalyzes the hydrolysis of polypeptide chains at the carboxyl side of the aromatic amino acids.
iii. **Bond specificity:** Enzymes act on substrate that are similar in structure and have same type of bond. For example:
    (a) *α-Amylase* hydrolyzes alpha-bonds of polysaccharides such as starch and glycogen.
    (b) *Lipases* hydrolyze ester bonds in different triglycerides.
iv. **Optical or stereospecificity:** Many enzymes exhibit stereospecificity, that is, enzymes are not specific only to the substrate but also to its optical isomer. This specificity is described as the highest specificity shown by an enzyme. Some of the examples in this category are:
    (a) *Fumarase* catalyzes the addition of water to fumarate to yield malate.
    (b) The hydrolysis of L-arginine to ornithine and urea is carried out by *arginase*.
    (c) *Glycerol kinase* catalyzes the conversion of glycerol to L-phosphoglycerol.

Enzyme specificity is essential for the normal functioning of cells as it regulates the metabolic pathways and prevents the unwanted side reactions at a particular active site. Enzymes showing highest specificity and accuracy also exhibit "proofreading" mechanisms and are involved in important biological processes like replication and gene expression. For example, DNA polymerase performs various crucial functions: it catalyzes the synthesis of DNA, checks the product, and also repairs it, if found any incorrect base pair.

3.3 Mechanism of Enzyme Action

In an enzyme-catalyzed reaction, the enzyme [E] binds to the substrate [S] to form a complex. The enzyme–substrate [ES] complex then forms the product [P]:

$$E + S \rightleftharpoons ES \rightleftharpoons P + E$$

A substrate usually binds to the small region of the enzyme called active site. The active site is the region within the enzyme where the substrate molecule binds, undergoes chemical reaction, and releases products. It is usually a groove or cleft in the enzyme formed by the folding pattern of the protein. As most of the enzymes are proteins, and proteins are made up of amino acids, the amino acid side chains align themselves in a manner that they can bind to the substrate via noncovalent interactions such as hydrogen bonds, ionic interactions, hydrophobic interactions, and van der Waals interactions. The reactive side chains of aspartate, glutamate, cysteine, lysine, arginine, serine, threonine, histidine, and hydrophobic residues play a crucial role in substrate binding. The substrate bound to the enzyme undergoes a conformational change, attaining a temporary tense state from which it transformed into products. Although the active site is a small portion of the enzyme involving only few amino acids in the catalysis process, it is the complete three-dimensional structure of protein which contributes to maintain a proper configuration at the catalytic site. The catalyzed reaction occurs at the active site of an enzyme in various steps as illustrated in Figure 3.1:

(i)   The first step involves the binding of a substrate to the active site of the enzyme.
(ii) In the second step, transition state is formed, bond rearrangement takes place and the process of catalysis occurs.
(iii) The process of bond-breaking and bond-making takes place in the third step, leading to the formation of products from substrates.
(iv) In the last step, the product is released from the enzyme when the reaction is complete (they no longer fit well in the active site). The free enzyme is now ready to catalyze the next set of reaction to form products.

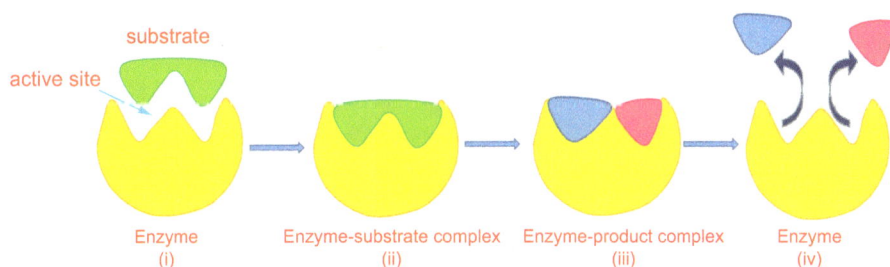

Figure 3.1: Steps involved in an enzyme-catalyzed reaction.

To facilitate any biochemical reaction, the reactant molecules must possess sufficient energy to cross a potential energy barrier, the activation energy (E_a). Activation energy is the minimum amount of energy required to bring a molecule to a state in which they can undergo a chemical reaction. The activation energy barrier to a biochemical reaction is lowered by the enzyme. All molecules possess varying amount of energy depending on their collision history but, generally, only a few have sufficient energy for the reaction. The lower the potential energy barrier to the reaction, the more reactants have sufficient energy and, hence, the faster the reaction will occur. All catalysts, including enzymes, function by forming a transition state, and in an enzyme-catalyzed reaction, transition state is formed by the combination of enzyme and substrate [ES]. This ES complex lowers the activation energy of the biochemical reaction and promotes its rapid progression. Figure 3.2 shows the reaction energy diagram showing enzyme-catalyzed and uncatalyzed reactions.

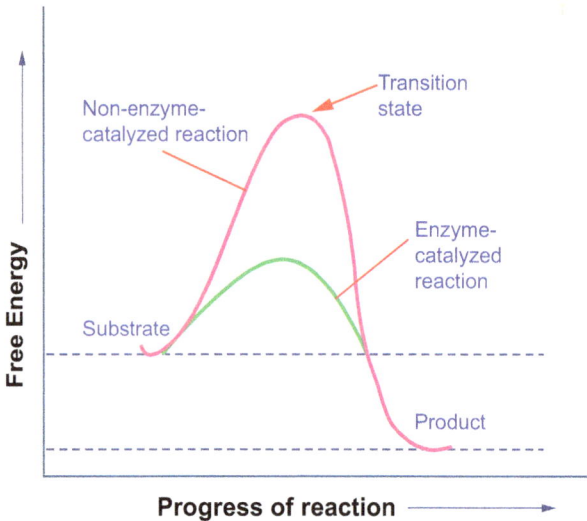

Figure 3.2: Reaction energy diagram showing enzyme-catalyzed and uncatalyzed reaction.

Two important theories have been developed to explain the mechanism of the enzyme action.

3.3.1 Lock-and-Key Model

This model was proposed by the German biochemist Emil Fischer in 1894 to explain the interaction of enzyme and substrate in an enzyme-catalyzed reaction. The theory behind this model is the complementarity between the active site of an enzyme and shape of the substrate. According to this model, both the enzyme and substrate possess fixed conformations which allow them to fit easily into one another and form an enzyme–substrate complex. This is similar as a key fit into a lock to open it. Each enzyme can accept one or two substrates and represents a "lock," whereas the substrates work like keys for the enzymes. The enzyme binds to the appropriate substrate, forms an enzyme–substrate complex, and catalyzes the chemical reaction by which products are released from the active site, as shown in Figure 3.3. The enzyme will not bind to the substrate if it has a different shape, although this model explains the remarkable specificity of enzymes but does not support the flexible side of the enzyme.

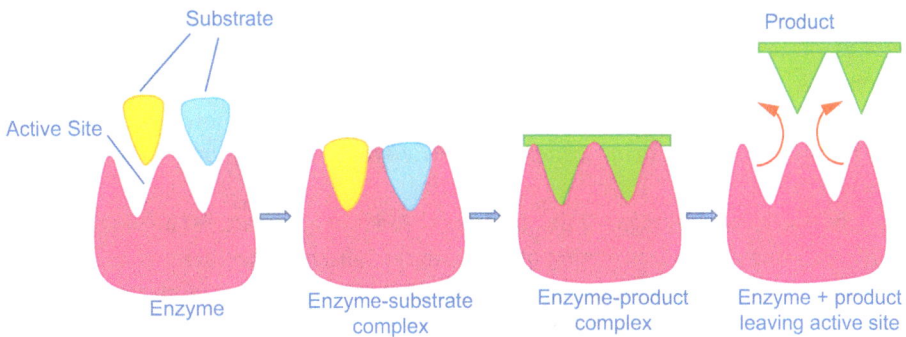

Figure 3.3: Lock-and-key model of the enzyme-catalyzed reaction.

3.3.2 Induced Fit Model

The lock-and-key model explains enzyme specificity, but it could not explain the stabilization of the transition state that enzymes achieve. The theory was accepted for many years, till Daniel Koshland's induced fit hypothesis was proposed in 1958. Koshland suggested a modified version of lock-and-key model that considers the conformational flexibility of enzymes. According to this model, the shape of the active site changes when a substrate molecule binds to such enzymes. The binding of the substrate induces a conformational change in the enzyme that results in a complementary fit after the substrate is bound, which means that the substrate does not simply binds in the rigid active site; the amino acid side chains that make up the active site configured themselves into the precise positions that enable the catalytic functioning of the enzyme. Figure 3.4 demonstrates the induced fit model of an enzyme-catalyzed reaction.

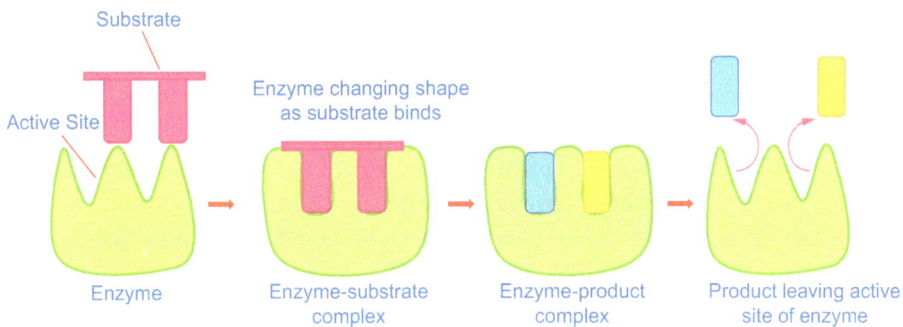

Figure 3.4: Induced fit model of enzyme-catalyzed reaction.

An example of induced fit mechanism is the binding of glucose (substrate) to the enzyme *hexokinase,* which induces a conformational change in the structure of the enzyme. As a result, the active site adopts a shape that is complementary to the substrate (glucose) and catalyzes the reaction. Different enzymes show features of both the models, some acquiring complementarity while others adopt conformational changes.

The spatial arrangement of the amino acid residues at the active site of an enzyme is the deciding factor for binding of a substrate. Substrate specificity is often determined by changing few amino acids at the active site and this can be clearly seen in a family of enzymes called *serine proteases (trypsin, chymotrypsin, and elastase).* These three enzymes catalyze the hydrolysis of peptide bonds in proteins and contain an essential serine residue at the active site which is important for catalysis. *Chymotrypsin* selectively cleaves the peptide bonds on the carboxyl terminal side of the large aromatic and hydrophobic amino acids such as tryptophan, tyrosine, and phenylalanine; the carboxyl side of positively charged lysine or arginine residues is cleaved by *trypsin;* and *elastase* cleaves majorly on the carboxyl side of residues with small uncharged side chains. Here, we are discussing the mechanism of action of *chymotrypsin.*

3.3.3 Chymotrypsin: Mechanism of Enzyme Action

Chymotrypsin is a digestive enzyme that plays a key role in the breakdown of polypeptides and proteins. It is synthesized as a single polypeptide as an inactivated precursor called *chymotrypsinogen,* which is then activated by *trypsin*-induced cleavage into three chains. The side chain of serine-195 is linked via hydrogen bond to the imidazole ring of histidine-57. In turn, the -NH group of this imidazole ring is hydrogen bonded to the carboxylate group of aspartate-102. This constellation of residues (His-57, Asp-102, and Ser-195) forms the catalytic triad, shown in Figure 3.5. The various steps involved in the hydrolysis of peptide bond, catalyzed by chymotrypsin, are:

Step 1: When a substrate (polypeptide) approaches and binds to the active site of the enzyme, the side chain of the residue next to the peptide bond to be cleaved resides in the hydrophobic pocket of the enzyme, positioning the peptide bond for attack. An alkoxide ion is produced via abstraction of one proton from serine by histidine. The resultant serine reacts with the substrate.

Step 2: In chymotrypsin, the carboxylate group of Asp-102 forms a hydrogen bond with -NH group of His-57. This results in compression of hydrogen bond and shifts electron density to the other nitrogen atom (not involved in the hydrogen bonding) of His-57. In doing so, the histidine becomes a strong base, which allows His-57 to deprotonate Ser-195 and turns it into a strong nucleophile that can attack the carbonyl carbon of peptide chain (substrate). There are now four atoms bonded to the carbonyl carbon, leading to the formation of an unstable tetrahedral intermediate in which oxygen atom bears a

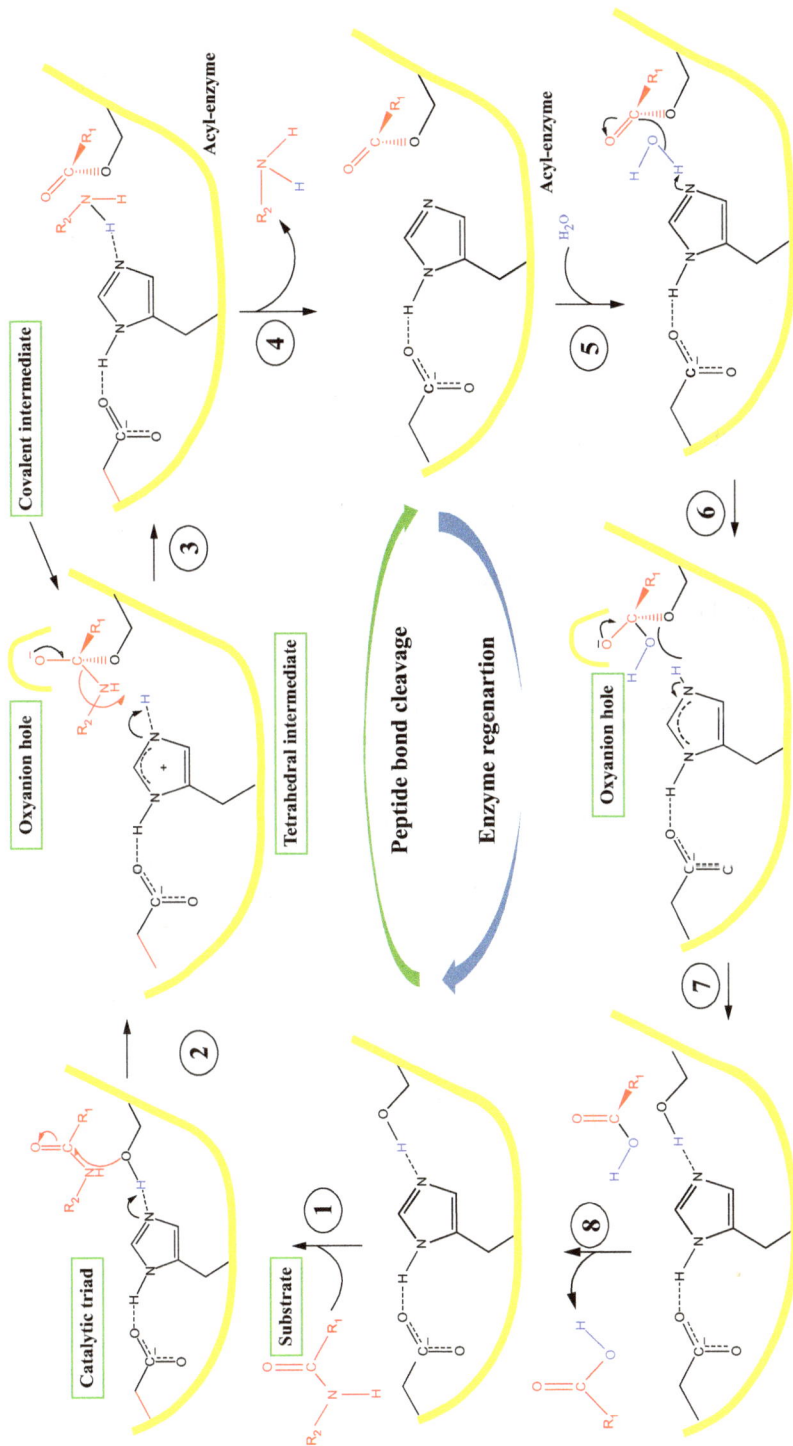

Figure 3.5: Peptide hydrolysis by chymotrypsin.

negative charge. This charge is stabilized by interactions with the -NH groups from the protein at a site called *oxyanion hole.*

Step 3: The tetrahedral intermediate then collapses to generate the acyl-enzyme. The amino-leaving group is protonated by His-57, facilitating its displacement.

Step 4: The amine component departs from the enzyme (metabolized by the body), completing the first stage of the hydrolytic reaction (acylation of enzyme). An acyl-enzyme intermediate can be observed by trapping them by adjusting the conditions such as temperature, pH, and nature of substrates by X-ray crystallography.

Step 5: In the next step, addition of water molecule takes place, which goes to the place that was earlier occupied by the amine component of the substrate.

Step 6: Histidine deprotonates the water to form a hydroxide ion (OH^-) which then attacks the carbonyl carbon atom of the acyl group and generates a second tetrahedral intermediate.

Step 7: Collapse of the tetrahedral intermediate leads to the formation of carboxylic acid product. The proton from histidine goes back to serine.

Step 8: The last step involves the release of the carboxylic acid, and the enzyme is reformed to catalyze the next reaction.

3.4 Factors Affecting Enzyme Activity

The activity of an enzyme is affected by its environmental conditions. Any change in the solution conditions alters the rate of reaction and hence the progress of an enzyme-catalyzed reaction. In nature, organisms adjust the conditions of their enzymes to produce an optimum rate of reaction, where necessary, or they may have enzymes that are adapted to function well in extreme conditions where they live. There are several factors which affect the rate at which enzymatic reactions proceed such as pH, temperature, ionic strength, enzyme concentration, substrate concentration, and the presence of any activators or inhibitors.

3.4.1 pH – Acidity and Basicity

pH refers to the hydrogen ion (H^+) concentration in solution, and it measures the acidity and basicity of a solution. The pH scale ranges from 0 to 14; however, negative pH values and values above 14 are also possible. Lower pH values mean higher H^+ concentrations and lower OH^- (hydroxide ion) concentrations.

The activity of an enzyme is its ability to function, and it can be defined as the micromoles of product formed per minute under standard conditions:

$$1\,\text{unit (U)} = 1\,\mu\,\text{mol/min}$$

The activity of an enzyme varies with the pH of the solution. As the pH of the solution is lowered, an enzyme will gain a proton (H^+); similarly the enzyme will lose a proton on raising the pH of the solution. Under both conditions, the ionization state of acidic and basic amino acids will be changed, which can interfere with the hydrogen bonding and ionic interactions that help in maintaining the tertiary structure of protein molecules. This interference causes a change in shape of the enzyme, and most importantly disrupts its active site. Any change in the shape or polarity of the active site would minimize the interaction between the substrate and enzyme, and hence affects the rate of biochemical reaction. Each enzyme has an optimum pH – pH at which the activity of an enzyme is maximum, and at this pH, the shape of the active site is best suited to the shape of the substrate. Any change in pH above or below the optimum pH will quickly cause a decrease in the rate of reaction as enzyme molecules will have active sites whose shapes are not, or less, complementary to the shape of their substrate. The effect of pH on the rate of an enzyme-catalyzed reaction is displayed in Figure 3.6.

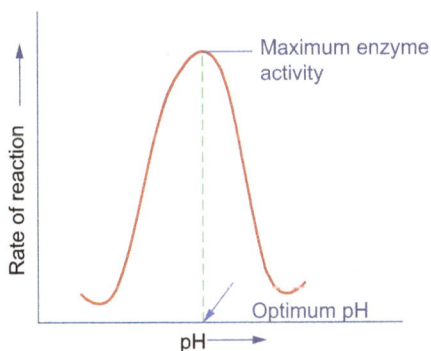

Figure 3.6: Effect of pH on the rate of reaction.

It is important to mention here that since pH is a logarithm function, even a change of one unit on the pH scale makes 10-fold difference in H^+ concentration. So, a small change in pH can affect the enzyme activity to a greater extent. But extreme changes in pH can cause the bonds to break, resulting in denaturation of enzyme and hence a complete loss of enzymatic activity. Enzymes present in different locations have different optimum pH as their environmental conditions are different (Table 3.2). Most of the enzymes present in the living organisms have an optimum pH in the range of 5.0–8.0 but they catalyze reactions most efficiently at neutral pH. However, there are some enzymes that function best at acidic or basic pH. For example, the digestive enzyme pepsin, found in the stomach, has an optimum pH at around 2.0.

Table 3.2: Enzymes and their optimum pH.

Enzyme	Optimum pH
Amylase (pancreas)	6.7–7.0
Catalase	7.0
Lipase (castor oil)	4.7
Lipase (pancreas)	8.0
Lipase (stomach)	4.0–5.0
Maltase	6.1–6.8
Pepsin	1.8–2.0
Trypsin	7.8–8.7
Urease	7.0

3.4.2 Temperature

As discussed earlier, the activity of an enzyme is affected by various factors, and temperature is one of the factors that affect the rate of biological reactions. We know that with the increasing temperature the kinetic energy of molecules gets increased causing them to move more rapidly, which enhances the chances of molecular collisions between them. As a result, like most chemical reactions, on increasing the temperature, the rate of an enzyme-catalyzed reaction also increases, forming more of the product. However, increasing the temperature also increases the vibrational energy of the atoms which makes up the enzyme molecules, and it can put a strain on various bonds that hold them together, especially the weaker hydrogen and ionic bonds. The bonds will break as a result of this strain, and hence will disrupt the active site of the enzyme. Any change in the shape of the active site will affect its binding with the substrate, and eventually the rate of enzyme-catalyzed reaction decreases. At extreme temperatures, the enzyme will become denatured and will lose its biological activity.

Similar to the optimum pH, each enzyme has a characteristic temperature at which its activity is maximum, which is known as optimum temperature and in humans, that temperature is around 37 °C – the normal body temperature. If the temperature is raised beyond the optimum temperature, the rate of enzyme-catalyzed reaction decreases due to the denaturation of enzyme. It is important to mention here that initially the rate of reaction will increase as temperature increases because of increased kinetic energy. However, with continuous rise in temperature, the effect of bond breaking will become greater and greater, and the rate of reaction will begin to decrease. The effect of temperature on the rate of an enzyme-catalyzed reaction is illustrated in Figure 3.7.

Most of the enzymes become denatured at temperature above 55 °C and stop working; however, there are certain organisms known as extremophiles that contain enzymes which function well even under extreme conditions. These enzymes are known as extremozymes, and they function under highly alkaline conditions, at high salt levels, high temperatures (thermophilic enzymes), and under other extreme conditions (pressure,

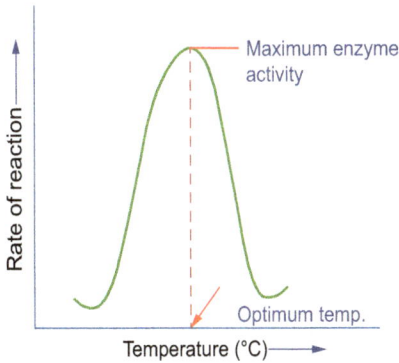

Figure 3.7: Effect of temperature on the rate of reaction.

acidity, etc.). Thermophile (archaea, some species of bacteria) is a type of extremophile that occurs in geothermally heated regions such as hot springs and peat bogs, and they contain enzymes that function at high temperatures. Thermophilic enzymes are used in biocatalysis, biodegradation, and biotransformation due to their extreme stability at elevated high temperatures. These enzymes also offer major applications in food, pulp, feeds, pharmaceutical, textile, detergents, and waste management industries.

3.4.3 Enzyme concentration

Like pH and temperature effect, changing the enzyme and substrate concentration also affect the rate of an enzyme-catalyzed reaction. Enzymes can catalyze a reaction even at a low concentration. If the concentration of the substrate is maintained constant, the rate of reaction will increase as the concentration of enzyme is increased. This is because of the presence of more active sites available for the binding of substrate molecules. However, this increase in enzyme activity does not occur forever, and this will only have an effect up to a certain concentration, after which any further increase in enzyme concentration would not increase the rate of reaction (Figure 3.8(a)). It is important to mention

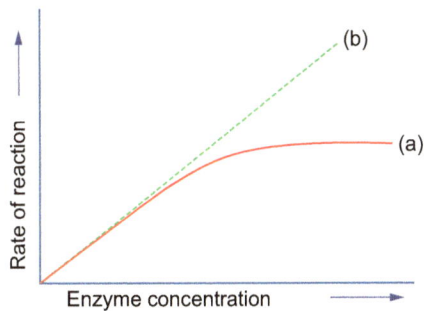

Figure 3.8: Effect of enzyme concentration on the rate of reaction.

that if the substrate is present in an excess amount, then the line (first part of the graph) can be approximated to a straight line Figure 3.8(b)).

3.4.4 Substrate concentration

In order to study the effect of substrate concentration on the rate of reaction, the concentration of enzyme must be kept constant. On gradually increasing the substrate concentration, the rate of reaction will increase until it reaches a maximum. This is because more substrate molecules will be interacting with the enzyme molecules resulting in more product formation. However, after a certain concentration, any further increase in substrate concentration will have no effect on the rate of reaction as the enzymes become saturated with substrate. A typical curve showing the variation of the reaction rate with an increase in the concentration of the substrate is shown in Figure 3.9.

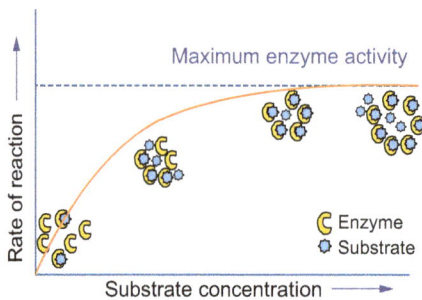

Figure 3.9: Effect of substrate concentration on the rate of reaction.

3.5 Cofactors and Coenzymes

Some enzymes require no chemical group other than their amino acid residues for their activity, and these enzymes are called simple enzymes, for example, *urease, pepsin, and trypsin*. There are some enzymes which require an additional small, nonprotein component to carry out a particular biochemical reaction. This additional group is known as cofactor and can be either one or more inorganic metal ions such as Zn^{2+}, Fe^{2+}, and Mg^{2+}, or a complex organic molecule composed of vitamins or vitamin derivatives such as NAD^+, $NADP^+$, and FAD (Flavin adenine dinucleotide), termed as coenzymes. Examples of some cofactor-bound enzymes are given in Table 3.3. Coenzymes play a significant role as a carrier of electrons and are involved in oxidation–reduction reactions occurring in the cell. A coenzyme or metal ion that is covalently bound to the enzyme is called a prosthetic group. The protein part of the enzyme without its cofactor is called apoenzyme, whereas a complete catalytically active enzyme along with its bound coenzyme or metal ion is called a holoenzyme. The structure of a holoenzyme consisting of apoenzyme and cofactor is shown in Figure 3.10.

Table 3.3: Examples of cofactor–bound enzymes.

Cofactor	Enzyme
Coenzyme	
Thiamine pyrophosphate	*Pyruvate dehydrogenase*
Flavin adenine nucleotide	*Monoamine oxidase*
Nicotinamide adenine dinucleotide	*Lactate dehydrogenase*
Coenzyme A (CoA)	*Acetyl CoA carboxylase*
Tetrahydrofolate	*Thymidylate synthase*
Metal	
Zn^{2+}	*Carbonic anhydrase*
Zn^{2+}	*Carboxypeptidase*
Mg^{2+}	*Hexokinase*
Mo	*Nitrogenase*
Ni^{2+}	*Urease*

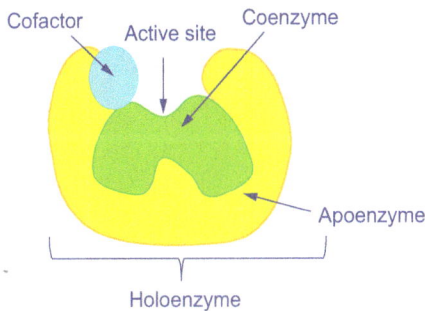

Figure 3.10: Apoenzyme and cofactor combines to form a holoenzyme.

The two important coenzymes, NAD^+ and FAD, involved in redox reactions are discussed further.

NAD^+ and $NADP^+$ are two important coenzymes that are prominent carriers of high-energy electrons. Their structure consists of the base adenine, two ribose sugars linked by phosphate groups, and a nicotinamide ring (Figure 3.11). The difference between $NADP^+$ and NAD^+ is that $NADP^+$ contains additional phosphate group joined to one of the ribose sugars.

The reactive part of both molecules is a nicotinamide ring which exists in either an oxidized or a reduced form. The reduced forms of NAD^+ and $NADP^+$ are NADH and NADPH, respectively. Both NAD^+ and $NADP^+$ act as carriers of electrons and are involved in oxidation–reduction reactions in various metabolic pathways. $NADP^+$ is used in anabolic (biosynthetic) reactions, while NAD^+ is more commonly used in

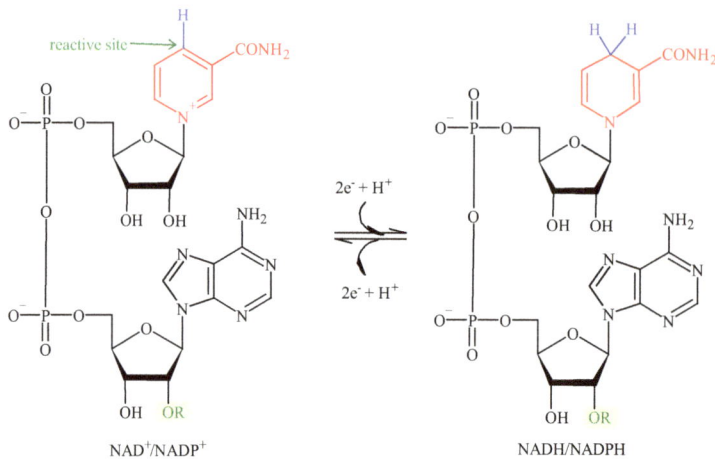

Figure 3.11: The structures of the coenzymes NAD$^+$ (NADH) and NADP$^+$ (NADPH). In NAD$^+$, R = H; in NADP$^+$, R = PO$_3^{2-}$.

catabolic (breakdown) reactions. In the oxidation of a substrate, the nicotinamide ring of NAD$^+$ accepts a hydrogen ion and two electrons, which are equivalent to a hydride ion (H:$^-$). For example,

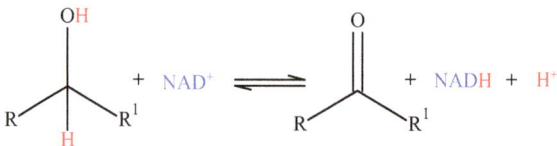

The other major electron carrier in the oxidation reaction is the coenzyme flavin adenine dinucleotide (FAD). This coenzyme exists in two different redox states, with FAD and FADH$_2$ being the oxidized and reduced forms, respectively. The structure of FAD consists of a flavin mononucleotide unit and an adenosine monophosphate unit (Figure 3.12), and the reactive part of FAD is its isoalloxazine ring. FAD can be reduced to FADH$_2$ by the addition of two electrons and two protons.

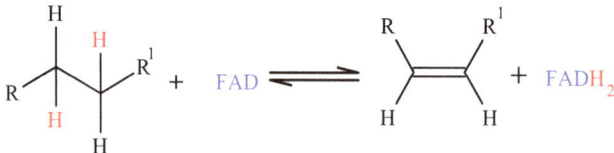

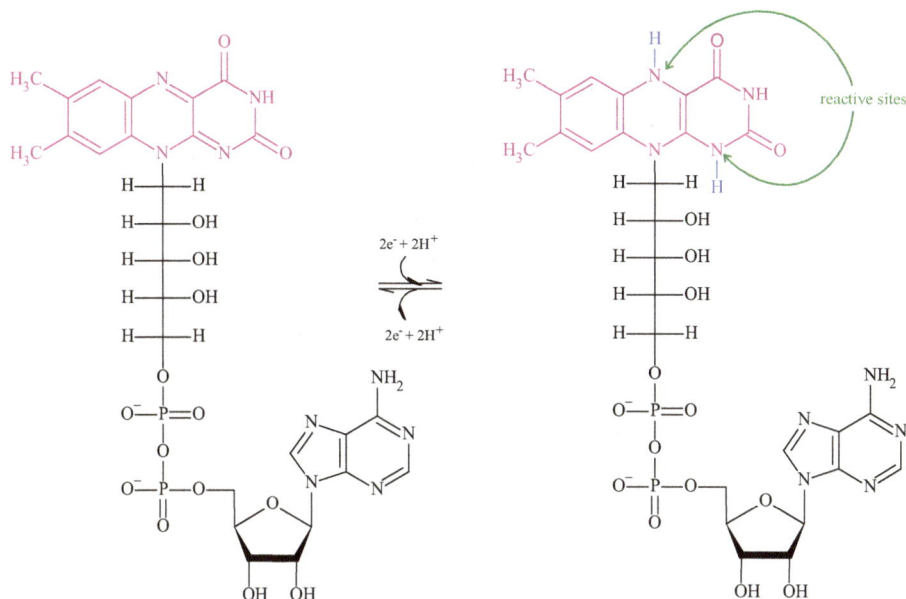

Figure 3.12: Structures of the coenzymes FAD and FADH$_2$.

3.6 Enzyme Inhibitors

The activity of an enzyme can be inhibited by the binding of different molecules or ions which can stop a substrate from entering the enzyme's active site. Any molecule which interferes with the activity of an enzyme and lowers its catalytic rate is called an inhibitor. Some enzyme inhibitors are normal metabolites which can correct a metabolic imbalance and regulate major controlled mechanisms in biological systems. In addition, many foreign molecules, such as drugs and toxin agents (poisons), are inhibitors of enzymes in the nervous systems. They block the enzymatic activity but usually do not destroy it. They are usually specific and generally work at low concentrations. Enzyme inhibitors can also be used as pesticides and herbicides. It is not necessary that all the molecules which bind to enzymes are inhibitors; there are some molecules that bind to enzymes and increase their enzymatic activity and these are termed as enzyme activators.

Enzyme inhibition can be of two types: reversible or irreversible. Irreversible inhibitors bind to the enzyme and often form a covalent bond. Reversible inhibitors, in contrast to irreversible inhibitors, bind noncovalently to the enzyme and can be classified as competitive, noncompetitive, and uncompetitive depending on whether these inhibitors bind to the enzyme, the enzyme–substrate complex, or both.

3.6.1 Reversible Inhibition

3.6.1.1 Competitive Inhibition

In the type of reversible inhibition called competitive inhibition, an inhibitor has close structural similarities to the substrate and it competes with the substrate molecule for the active site, thus preventing it from binding to the same active site, as illustrated in Figure 3.13. An enzyme may bind either to the substrate molecule, forming an enzyme–substrate (ES) complex or inhibitor, forming an enzyme–inhibitor (EI) complex but not both, at the same time, ESI, enzyme–substrate–inhibitor complex. A competitive inhibitor binds reversibly to the active site and decreases the rate of catalysis by reducing the number of enzyme molecules bound to the substrate; however, its effect can be reversed by increasing the substrate concentration. At high substrate concentration, it can successfully compete out the inhibitor in binding to the active site.

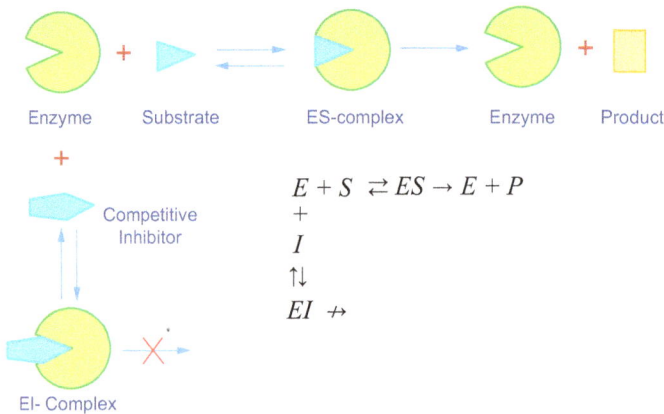

$$E + S \rightleftarrows ES \rightarrow E + P$$
$$+$$
$$I$$
$$\uparrow\downarrow$$
$$EI \nrightarrow$$

Figure 3.13: The characteristics of competitive inhibition.

Competitive inhibitors are commonly used as drugs. A good example of competitive inhibition is methotrexate, a structural analog of dihydrofolate, which is a potent competitive inhibitor of the enzyme *dihydrofolate reductase* (DHFR) and is used to treat cancer (Figure 3.14). DHFR catalyzes the conversion of dihydrofolate to the active tetrahydrofolate which is essential for the biosynthesis of pyrimidines and purines. Methotrexate binds to DHFR and hence inhibits the synthesis of nucleic acids, and the affinity of methotrexate for DHFR is 1,000 times fold that of dihydrofolate (natural substrate).

Another example which belongs to this class is malonate, which is a competitive inhibitor of the enzyme *succinate dehydrogenase* (SDH). *Succinate dehydrogenase* plays a vital role in citric acid cycle (CAC) and electron transport chain. SDH catalyzes the oxidation of succinate to fumarate in the CAC. Malonate binds to the active site of the enzyme and hence competes with the enzyme's normal substrate, that is, succinate, and inhibits the formation of fumarate (Figure 3.15). Statins, also known as

(a)

Dihydrofolate

Methotrexate

(b)

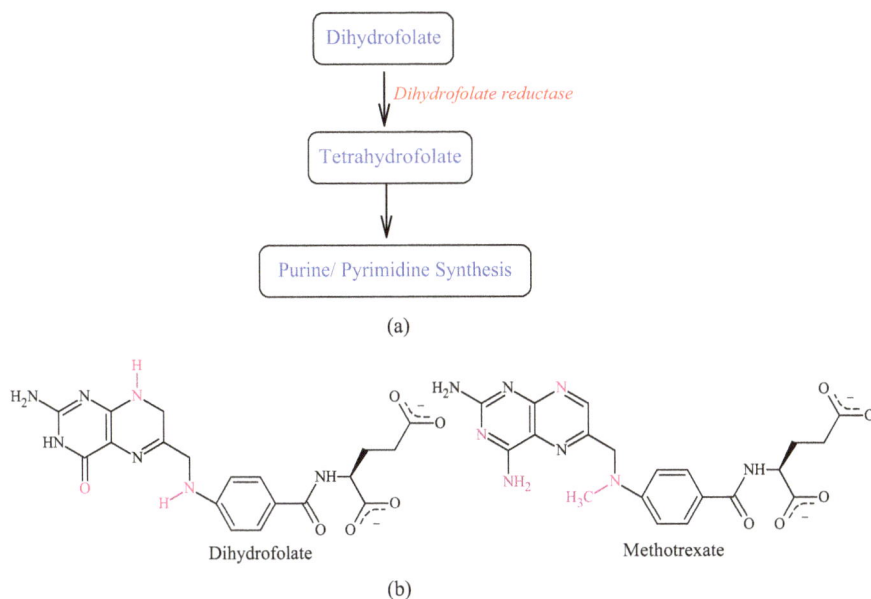

Figure 3.14: (a) Conversion of dihydrofolate to tetrahydrofolate catalyzed by *dihydrofolate reductase* and (b) the substrate dihydrofolate and its structural analog methotrexate.

HMG-CoA reductase inhibitors, are drugs which are employed to lower down the cholesterol level in the body. The enzyme *HMG-CoA reductase* controls the production of cholesterol in the liver, and inhibition of this enzyme slows down the biosynthesis of cholesterol.

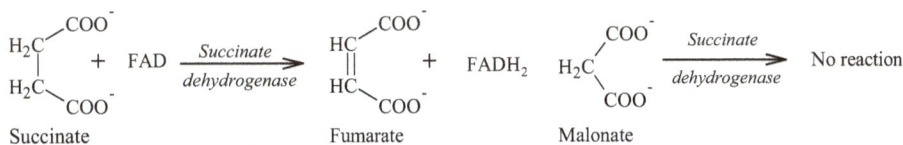

Succinate

Fumarate

Malonate

Figure 3.15: Inhibition of succinate dehydrogenase by competitive inhibitor malonate.

3.6.1.2 Noncompetitive Inhibition

A noncompetitive inhibitor has a structure different than that of a substrate and binds reversibly at a site other than the active site shown in Figure 3.16. Binding at some other site changes the three-dimensional shape of the enzyme, which then prevents binding of the substrate resulting in decreased efficacy of the enzyme. Since the inhibitor binds at a different site than that of a substrate, it can bind to free enzyme or to the enzyme–substrate complex. Unlike competitive inhibition, the effect of a noncompetitive inhibitor cannot be overcome by increasing the substrate concentration.

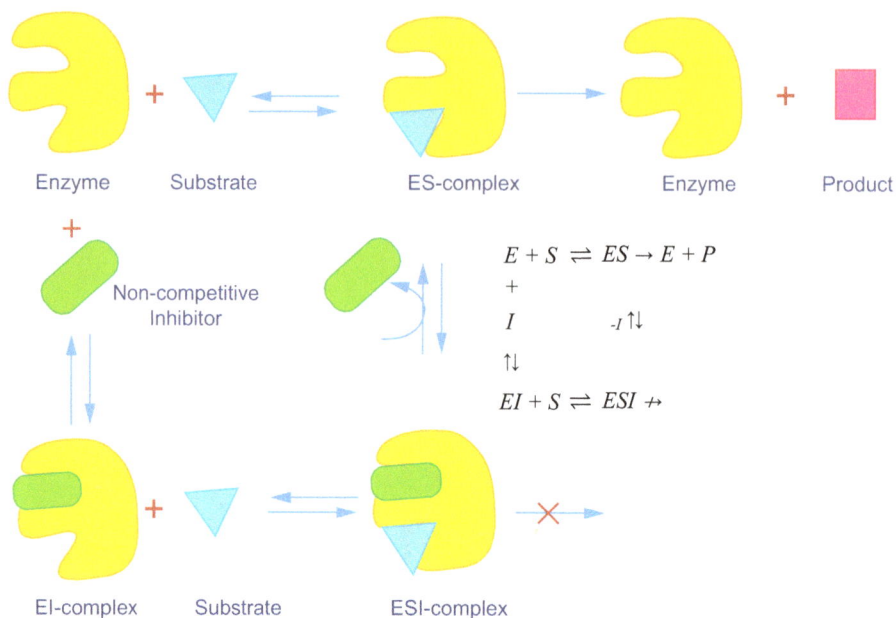

Figure 3.16: The characteristics of noncompetitive inhibition.

Noncompetitive inhibitors play an important role in regulation of metabolism via feedback mechanism. Glycolysis is one among various metabolic pathways that have been regulated by noncompetitive inhibition. For example, the amino acid alanine and ATP are noncompetitive inhibitors of the enzyme *pyruvate kinase* (PK), the enzyme that catalyzes the conversion of phosphoenol pyruvate to pyruvate (the final step of glycolysis). The inhibition of PK allows cells to shut off the breakdown of glucose (on acquiring adequate amounts of end products), thus preventing overproduction and wasting of cellular energy. Another example of noncompetitive inhibitor is glucose-6-phosphate, which is a potent noncompetitive inhibitor of *hexokinase*. *Hexokinase* catalyzes the first step of glycolysis which is the conversion of glucose into glucose-6-phosphate. The inhibition of *hexokinase* by glucose-6-phosphate leads to shutting off the process if large amount of glucose is already broken down during the glycolytic pathway. Doxycycline (an antibiotic) is another example in this category which at low concentrations acts as a noncompetitive inhibitor of matrix *metalloproteases* and is used to treat periodontal disease.

3.6.1.3 Uncompetitive Inhibition

In uncompetitive inhibition, the inhibitor specifically binds only to the enzyme–substrate complex, and this type of enzyme inhibition is relatively rare in comparison to competitive and noncompetitive inhibition. In this type of inhibition, the inhibitor does not have to bind at the active site rather the binding site for inhibitor is created upon interaction of the enzyme and substrate (Figure 3.17).

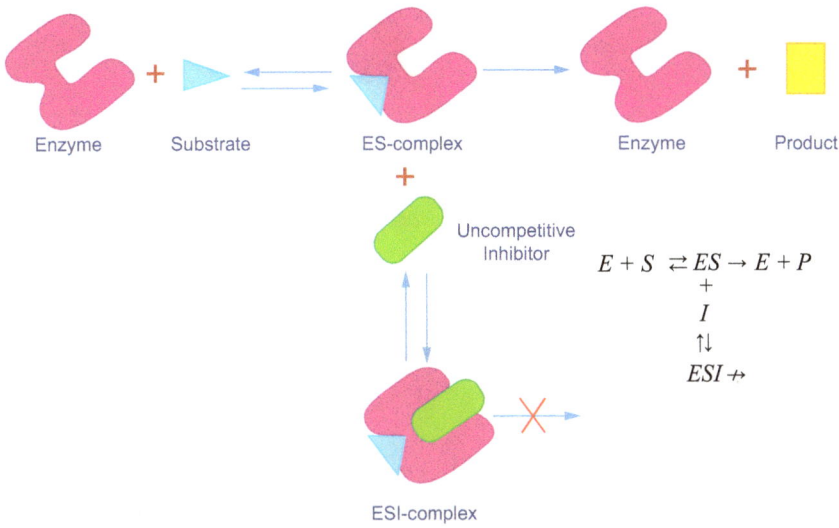

Figure 3.17: The characteristics of uncompetitive inhibition.

Glyphosate, also known as Roundup, is a widely used herbicide which acts as an uncompetitive inhibitor of *5-enolpyruvylshikimate 3-phosphate synthase*. This enzyme is only produced by plants and microorganisms, and catalyzes the biosynthesis of aromatic amino acids (tyrosine, tryptophan, and phenylalanine) via the shikimate pathway.

3.6.2 Irreversible Inhibition

Irreversible inhibitors bind covalently to the enzyme, modify key amino acid residues at or near active sites which are needed for enzymatic activity, and permanently inactivate the enzyme. Irreversible inhibitors can be categorized into group-specific reagents, substrate analogs (affinity labels), and suicide inhibitors. Some important drugs fall under the category of irreversible inhibitors.

3.6.2.1 Group-Specific Reagents

These inhibitors react with specific side chains of amino acids. Diisopropylphosphofluoridate (DIPF) and iodoacetamide are two examples of group-specific reagents. DIPF, a component of nerve gases, forms a covalent bond with the hydroxyl (-OH) group of serine residue at the active site of the enzyme *acetylcholinesterase*, thereby inhibiting the enzyme and preventing neurotransmission. DIPF also inactivates the proteolytic enzyme *chymotrypsin* by binding only 1 serine residue (serine 195) among 28 serine residues resulting in a complete loss of enzyme activity (Figure 3.18).

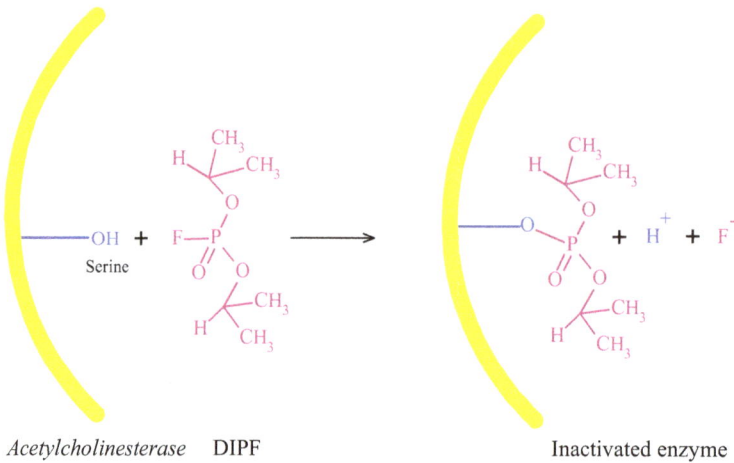

Acetylcholinesterase   DIPF   Inactivated enzyme

Figure 3.18: Inhibition of *acetylcholinesterase* by DIPF (diisopropylphosphofluoridate), a group-specific reagent.

The other example of group-specific reagent is iodoacetamide, which is an irreversible inhibitor of all cysteine peptidases. It binds covalently with the thiol (-SH) group of cysteine residues and prevents the formation of disulfide bonds as shown in Figure 3.19.

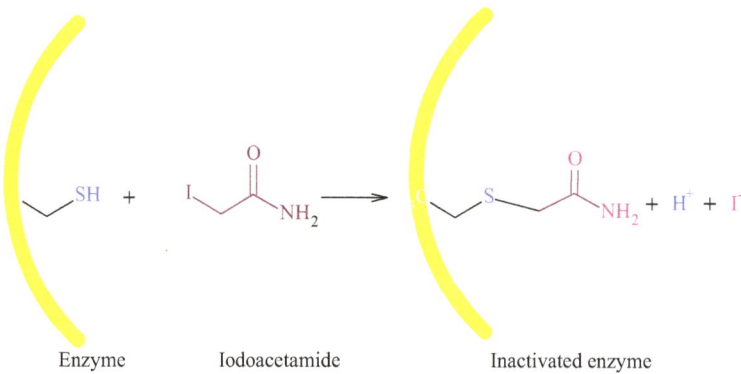

Enzyme   Iodoacetamide   Inactivated enzyme

Figure 3.19: Inhibition of enzyme by iodoacetamide, a group-specific reagent.

3.6.2.2 Affinity Labels

Affinity labels are irreversible enzyme inhibitors that have structural similarity to a particular substrate and bind covalently to their target causing their inactivation. They have more specificity for the active site of an enzyme than group-specific reagents. Examples in this category are tosyl-L-phenylalanyl chloromethyl ketone (TPCK) and bromoacetol phosphate.

TPCK binds at the active site of the enzyme *chymotrypsin* and modifies the histidine residue at that site, thus inhibiting the enzymic activity (Figure 3.20).

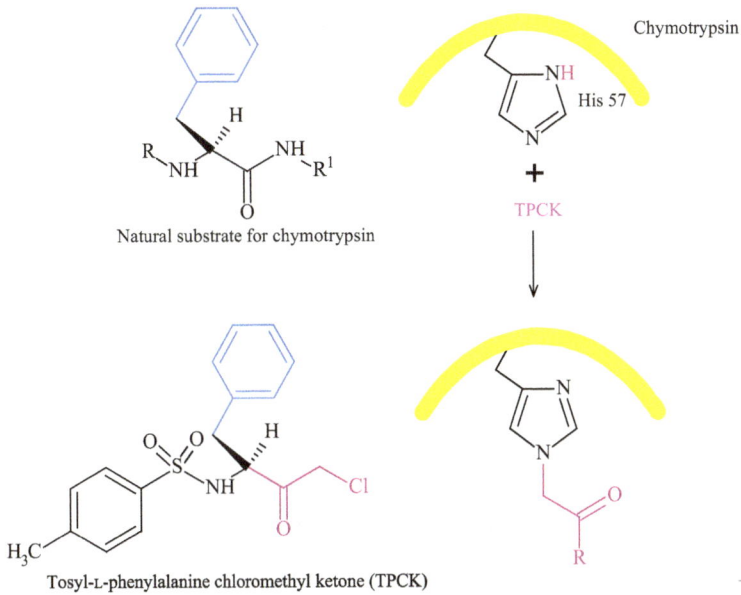

Figure 3.20: Inhibition of chymotrypsin by TPCK (tosyl-L-phenylalanine chloromethyl ketone).

An analog of dihydroxyacetone phosphate (DHAP), that is bromoacetol phosphate, binds at the active site of *triose phosphate isomerase* and then reacts irreversibly with the glutamic acid residue required for enzymic activity (Figure 3.21).

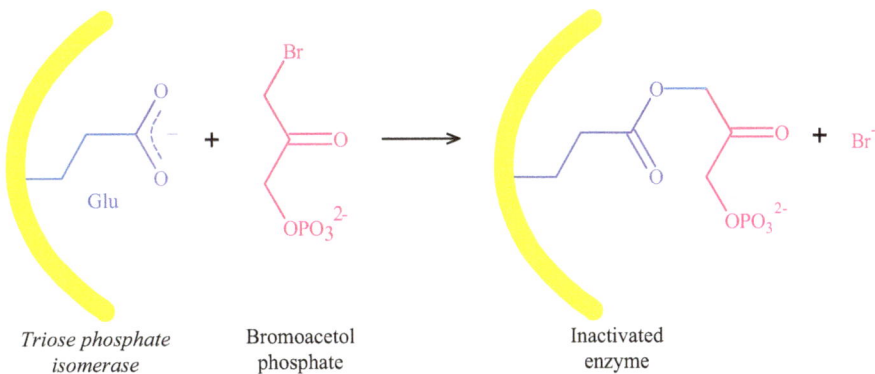

Figure 3.21: Inhibition of *triose phosphate isomerase* by bromoacetol phosphate, an analog of DHAP (dihydroxyacetone phosphate).

3.6.2.3 Suicide Inhibitors

Suicide inhibition, also known as mechanism-based inhibition, is an irreversible form of enzyme inhibition in which inhibitor binds at the active site of the enzyme and generates a reactive intermediate during normal catalysis process. The intermediate then reacts irreversibly with the enzyme to form a stable enzyme–inhibitor complex and inactivates the enzyme. For example, N,N-dimethylpropargylamine is an inhibitor of mitochondrial enzyme *monoamine oxidase* (MAO). MAOs (A and B) are FAD (cofactor)-dependent enzymes that play a central role in neurotransmitter metabolism, and altered MAO levels may associate with a number of psychiatric and neurological diseases. A flavin moiety of MAO oxidizes the N,N-dimethylpropargylamine, which in turn inactivates the enzyme by covalently modifying the N-5 of the flavin prosthetic group (Figure 3.22). The suicide inhibitors of MAO, N,N-dimethylpropargylamine, and (–)-deprenyl are used to treat depression and Parkinson's disease.

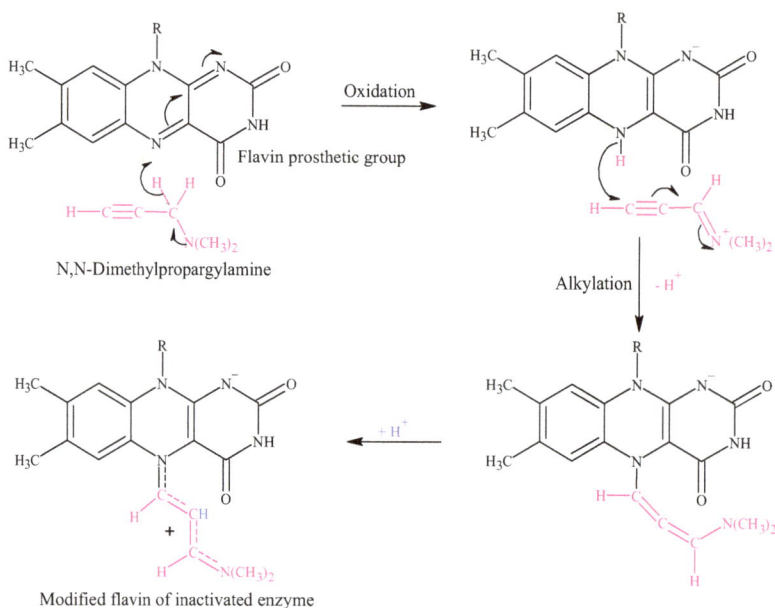

Figure 3.22: Inhibition of enzyme *monoamine oxidase* by N,N-dimethylpropargylamine.

The first antibiotic, penicillin, is another important example of suicide inhibitor. Penicillin covalently attaches to the serine residue at the active site of the enzyme *glycopeptide transpeptidase*, inhibiting its cross-linking activity, preventing the synthesis of bacterial cell walls, followed by killing the bacteria (Figure 3.23).

Enzyme inhibitors exhibit several therapeutic applications and play an important role in elucidating metabolic pathways in cells. Most of the drugs used are either competitive inhibitors or mechanism-based suicide inhibitors, and they inhibit enzymes,

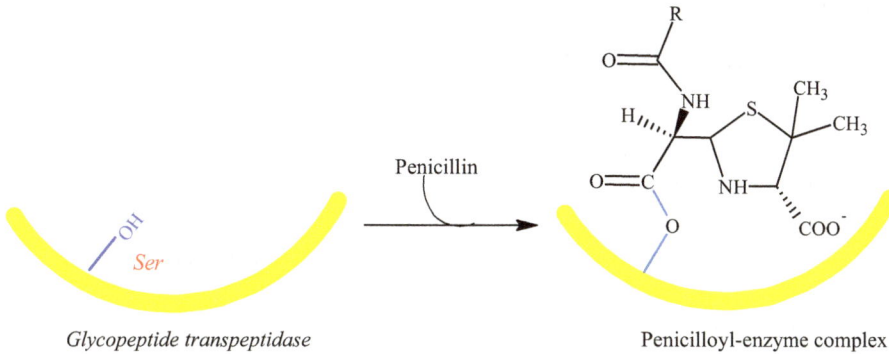

Figure 3.23: Inhibition of *glycopeptide transpeptidase* by penicillin.

thereby correcting a pathological condition. Inhibitors are also used to cure viral infections as they inhibit the viral enzymes which are responsible for the replication. They are also used as pesticides and herbicides. As discussed, many drugs act as enzyme inhibitors, and researchers are actively trying to discover new inhibitors and improve the efficacy of the existing ones so that they exhibit high specificity, low toxicity, and minimized side effects.

References

[1] Berg JM, Tymoczko JL, Gatto Jr GJ. Lubert Stryer. Biochemistry. Freeman Macmillan. 8th edition. ISBN-13. 2006:978–1.
[2] Bhutani SP. Chemistry of biomolecules. CRC Press. Boca Raton, Florida; 2019 Sep 25
[3] Hames D, Hooper N. Instant notes biochemistry. Taylor & Francis; 2006 Sep
[4] Laidler KJ. A brief history of enzyme kinetics. New Beer in an Old Bottle: Eduard Buchner and the Growth of Biochemical Knowledge, Valencia, Spain: Universitat de Valencia. 1997:127–33. Cornish-Bawden A, editor. New Beer in an Old Bottle. Eduard Buchner and the Growth of Biochemical Knowledge. Universitat de València; 1997.
[5] Naik P. Biochemistry. JP Medical Ltd India; 2015 Nov 30.
[6] Tipton, K. Translocases (EC 7): A new EC Class. Enzyme Nomenclature News, August 2018 Captions

Chapter 4
Concept of Energy in Biosystems

The highly structured and organized nature of living systems is perceptible and astonishing. Growth, development, and metabolism are some of the fundamental processes that occur in living organisms, and the role of energy is fundamental to all these biological processes. The survival of any living organism depends on energy transformations, that is, the exchange of energy within and without a particular system. The fundamental matter in bioenergetics, that is, the study of energy relationships and conversions in living organisms, signifies the way by which energy from fuel metabolism or by capturing light is coupled to the energy-requiring reactions occurring in the cell. Muscular contraction, synthetic reactions, and active transport are some of the important processes that get energy when linked or coupled with some energy-releasing reactions (exergonic reactions). In all organisms (autotrophic and heterotrophic), ATP (adenosine triphosphate) plays an important role in transferring energy from the exergonic to the endergonic reactions. ATP is called a high-energy phosphate compound and is produced by living organisms via oxidative phosphorylation. The terminal phosphate linkage in ATP is relatively weak; when broken, it yields adenosine monophosphate (AMP) and inorganic phosphate and releases a large amount of energy. An organism's stockpile of ATP is used by the cells to perform different activities to sustain life, and energy released from rearrangement of bonds within molecules is utilized to power all biological processes in every organism.

Bioenergetics or biochemical thermodynamics deals with the transformations, exchange, requirements, and processing of energy within living systems. It also focuses on how cells transfer energy. Some of the essential biological processes such as biosynthesis of nucleic acids and other biomolecules are not thermodynamically favored under provided conditions, as they require an input of energy. They can proceed if coupled with energy-releasing processes. So, it endows with the answer why some reactions may occur while others do not.

4.1 Metabolism

All cells require a continuous input of energy to sustain life. Living organisms need energy for physical movements, for active transport of ions and molecules, and for synthesis of biological molecules from their precursors. Our body uses energy, even during sleep, for various activities such as for pumping blood, breathing, supplying oxygen to tissues, maintaining body temperature, synthesizing new tissues for growth, and repairing the damaged ones. Where does this energy come from to perform all such activities? Phototrophs (photosynthetic organisms) get this energy by trapping sunlight and

https://doi.org/10.1515/9783110793765-004

making carbohydrates in a process called photosynthesis, whereas chemotrophs get energy through chemical processes obtained by the oxidation of foodstuffs made by phototrophs.

Metabolism refers to the linked series of chemical reactions that occur in living organisms providing energy to sustain life. The metabolic pathways are divided into two broad classes: catabolic and anabolic. The chemical reactions that require an input of energy – such as the synthesis of DNA, glucose, or fats from their simple precursors – are called anabolic reactions or, more generally, *anabolism*:

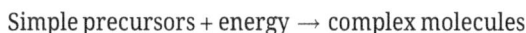

$$\text{Simple precursors} + \text{energy} \rightarrow \text{complex molecules}$$

The set of chemical reactions that breaks down a complex molecule such as carbohydrates and fats into simpler ones, together with the release of energy, are termed as catabolic reactions or *catabolism*:

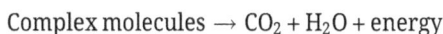

$$\text{Complex molecules} \rightarrow CO_2 + H_2O + \text{energy}$$

For example, when we eat rice or bread, starch gets converted into glucose units with the help of enzymes, and cells further catabolize these glucose molecules to release useful energy to perform mechanical work in cellular movements and muscle contraction. Conversely, glucose molecules can be united to form glycogen for storage via anabolic pathways. Some pathways can function both in anabolic and catabolic contexts, depending on the availability of energy in the cell and such pathways are termed as amphibolic pathways. For example, citric acid cycle (CAC) breaks down acetyl coenzyme A (CoA) and is associated with the synthesis of amino acids. Figure 4.1 illustrates catabolic and anabolic reactions.

4.2 Standard Free Energy Change in Biochemical Reactions

Now the question arises, how an individual reaction contributes toward specific pathways? A reaction pathway must fulfill two criteria: (i) the individual reactions should be specific, that is, it should give only one particular product (or products), and (ii) the series of reactions that constitute the pathways should be thermodynamically favored. The thermodynamics of metabolic pathways are most readily understood in terms of free energy, and laws of thermodynamics are of utmost importance as they govern the conditions under which a particular reaction will occur or not. These laws are general principles that govern all physical and biochemical processes and make a clear difference between system and surrounding. A system is the quantity of matter present within a defined region, whereas the matter in the rest of the universe constitutes the surrounding.

The first law of thermodynamics states that the total energy of a system and its surroundings is constant. In other words, energy content of the universe is constant; energy can neither be created nor destroyed. However, energy may be transferred

CATABOLIC REACTIONS

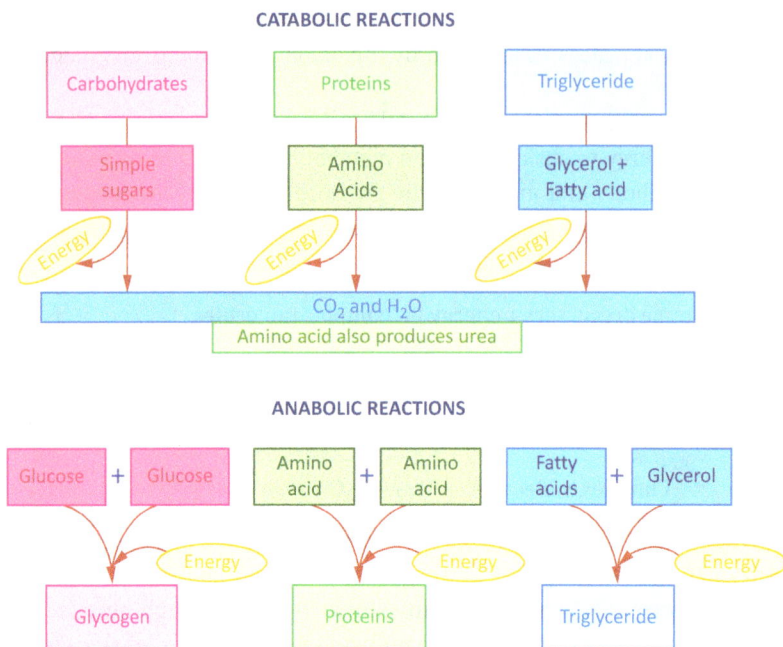

Figure 4.1 (upper portion):

Carbohydrates → Simple sugars

Proteins → Amino Acids

Triglyceride → Glycerol + Fatty acid

Energy

CO_2 and H_2O

Amino acid also produces urea

ANABOLIC REACTIONS

Glucose + Glucose → Glycogen (Energy)

Amino acid + Amino acid → Proteins (Energy)

Fatty acids + Glycerol → Triglyceride (Energy)

Figure 4.1: Catabolic and anabolic reactions.

from one form to another. For example, in living systems, chemical energy may be transformed into heat, electrical energy, radiant energy, or mechanical energy.

Entropy is another important thermodynamic concept and is associated with a state of randomness and disorder of a system. Any change in the randomness is expressed as ΔS, and positive value of ΔS indicates an increase in disorder/randomness. The second law of thermodynamics states that the total entropy of a system and its surrounding always increases for a spontaneous process. J. Willard Gibbs gave the theory of change in energy during a chemical reaction and showed that free energy content, G, of any closed system can be defined as follows:

$$G = H - TS$$

where H is the enthalpy (heat), S is the entropy, and T is the absolute temperature (in kelvin, K).

When a chemical reaction takes place at a constant temperature and pressure, the change in free energy (ΔG) of a reacting system is given by the following equation, which combines the two laws of thermodynamics:

$$\Delta G = \Delta H - T\Delta S$$

where ΔS is the change in entropy, ΔH is the change in enthalpy (heat), and T is the absolute temperature (in K).

For a reaction to be spontaneous, the free energy change should be negative, and a negative ΔG occurs when the overall entropy of the universe increases. On the other hand, if ΔG is positive, the reaction is nonspontaneous. The system is at equilibrium and no net change takes place when $\Delta G = 0$.

For a chemical reaction,

$$A + B \rightleftharpoons C + D$$

The ΔG of this reaction is given by

$$\Delta G = \Delta G^\circ + RT \ln \frac{[C][D]}{[A][B]} \tag{4.1}$$

where ΔG° is the standard free energy change, R is the gas constant, T is the absolute temperature, and $[A]$, $[B]$, $[C]$, and $[D]$ are the molar concentrations of the reactants. ΔG° is the free energy change under standard conditions – that is, when each of the reactants A, B, C, and D is present at a concentration of 1.0 M. For biochemical reactions, the standard state is defined as having a pH of 7.0. So, the standard free energy change at pH 7.0 is denoted by $\Delta G^{\circ\prime}$.

At equilibrium, $\Delta G = 0$; therefore, eq. (4.1) becomes

$$0 = \Delta G^{\circ\prime} + RT \ln \frac{[C][D]}{[A][B]} \tag{4.2}$$

and so

$$\Delta G^{\circ\prime} = -RT \ln \frac{[C][D]}{[A][B]} \tag{4.3}$$

The equilibrium constant under standard conditions K'_{eq} is defined as follows:

$$K'_{eq} = \frac{[C][D]}{[A][B]} \tag{4.4}$$

Substituting eq. (4.4) into (4.3),

$$\Delta G^{\circ\prime} = -RT \ln K'_{eq} \tag{4.5}$$

$$\Delta G^{\circ\prime} = -2.303 \, RT \log K'_{eq} \tag{4.6}$$

Thus, eq. (4.6) shows that a change in free energy of a reaction is related to the equilibrium constant, K'_{eq}. It is pertinent to mention here that the actual value of ΔG can be smaller or larger than $\Delta G^{\circ\prime}$, as it depends on the concentrations of the reactants, solvent, presence of various proteins, and ions.

Some of the important biological processes are thermodynamically unfavorable as they require energy to proceed, and to carry out such reactions, they are coupled with other reactions that release free energy. The energy required for such endergonic reactions is provided by the hydrolysis of phosphoanhydride bonds of energy-rich phosphate compounds such as ATP. The sum of the free energy changes is negative, and the overall process is exergonic. For a chemically coupled series of reactions, the overall free energy change is equal to the sum of the free energy change of the individual steps.

4.3 Exergonic and Endergonic Reactions

Exergonic reactions are those chemical reactions that release energy, and ΔG is negative for such reactions, that is, these reactions are thermodynamically favored. Symbolically, for any exergonic reaction,

$$\Delta G = G_{products} - G_{reactants} < 0$$

Endergonic reactions are anabolic reactions that require energy and have a positive ΔG value. These reactions are thermodynamically unfavorable. Under constant temperature and pressure, the change in the standard Gibbs free energy would be positive, $\Delta G > 0$. Figure 4.2 shows the diagrammatic representations of exergonic and endergonic reactions.

Examples of exergonic reactions are cellular respiration, and the breakdown of ATP to ADP (adenosine diphosphate). Photosynthesis and synthesis of proteins from amino acids are the processes that fall under the category of endergonic reactions.

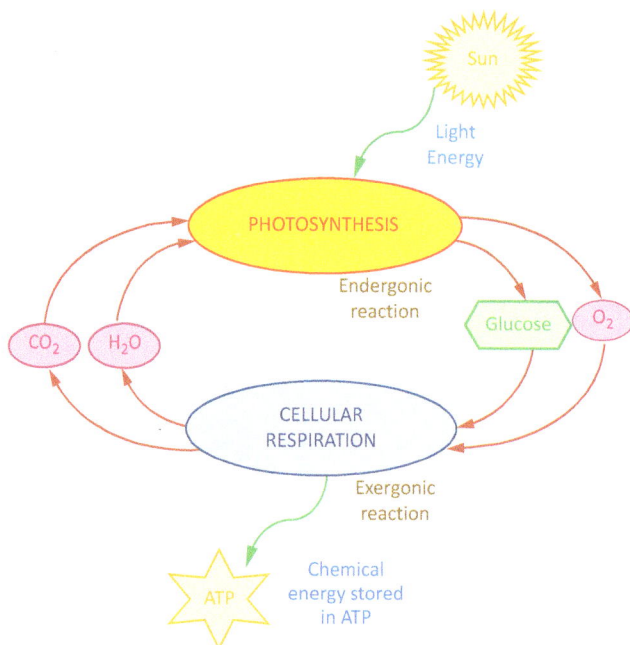

Figure 4.2: Representations of exergonic and endergonic reactions.

4.4 ATP: Universal Currency of Free Energy in Biological Systems

Biochemical reactions employed some common currencies as a medium for exchange of energy as most economies simplify trade by using currencies. The most important of the currencies in biochemical systems is ATP. ATP is an energy-rich molecule which consists of adenine (nitrogenous base), a ribose (pentose sugar), and triphosphate groups (Figure 4.3). ATP forms an active complex with magnesium (Mg^{2+}), and it contains two phosphoanhydride bonds, which when broken down releases a tremendous amount of energy that is used by the cells to power the biological work. ATP can be hydrolyzed to ADP (Adenosine diphosphate) and an inorganic phosphate, orthophosphate (P_i), or to AMP (Adenosine monophosphate) and pyrophosphate (PP_i). The reaction can proceed in either direction; ATP is formed from ADP and P_i, and exchange of energy in the biological systems takes place through this ATP–ADP cycle:

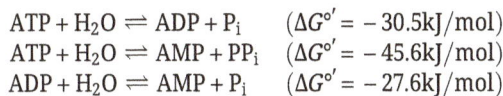

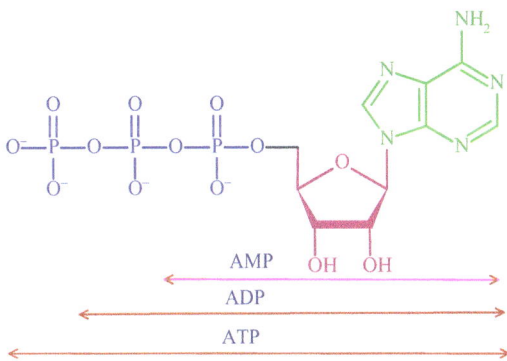

$$ATP + H_2O \rightleftharpoons ADP + P_i \quad (\Delta G^{o'} = -30.5 \text{kJ/mol})$$
$$ATP + H_2O \rightleftharpoons AMP + PP_i \quad (\Delta G^{o'} = -45.6 \text{kJ/mol})$$
$$ADP + H_2O \rightleftharpoons AMP + P_i \quad (\Delta G^{o'} = -27.6 \text{kJ/mol})$$

Figure 4.3: Structure of AMP, ADP, and ATP.

Some important biosynthetic reactions are carried out by the hydrolysis of other nucleoside triphosphates (NTPs) such as guanosine triphosphate (GTP), cytidine triphosphate, and uridine triphosphate. The diphosphate forms of these nucleotides are represented as guanosine diphosphate, cytidine diphosphate, and uridine diphosphate, and the monophosphate forms are denoted by guanosine monophosphate, cytidine monophosphate, and uridine monophosphate. NTPs can be synthesized by the phosphorylation of nucleoside diphosphates (NDPs), and this reaction is catalyzed by the enzyme *nucleoside diphosphate kinase.* The phosphorylation of nucleoside monophosphates (NMPs) is catalyzed by the enzyme *nucleoside monophosphate kinase:*

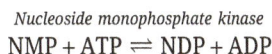

Nucleoside monophosphate kinase
$$NMP + ATP \rightleftharpoons NDP + ADP$$

Nucleoside diphosphate kinase
$$NDP + ATP \rightleftharpoons NTP + ADP$$

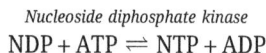

Like ATP, GTP also holds large amount of energy and supplies it for the synthesis of proteins and glucose (gluconeogenesis). It also plays an important role during the elongation stage in translation and in signal transduction. Although all NTPs are energetically equivalent, ATP is the primary carrier of cellular energy and plays a significant role in energy metabolism. When breaking down a complex molecule into simpler ones, metabolic reactions release high energy electrons, and further reactions transfer energy from these electrons to ATP. These high-energy electrons require special carriers to reach the ATP production site, and two main energy carriers are nicotinamide adenine dinucleotide (NAD^+) and flavin adenine dinucleotide (FAD) (discussed in Chapter 3).

How an exergonic reaction can be coupled with an endergonic reaction to drive a thermodynamically unfavorable reaction in living organisms?

This can be explained by taking the example of conversion of glucose to glucose 6-phosphate, the first step in glycolytic pathway. The reaction is

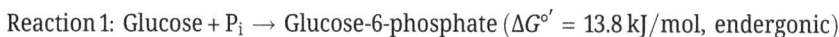

Reaction 1: Glucose + P_i → Glucose-6-phosphate ($\Delta G^{\circ\prime}$ = 13.8 kJ/mol, endergonic)

Reaction 2: ATP → ADP + P_i ($\Delta G^{\circ\prime}$ = − 30.5 kJ/mol, exergonic)

Reaction 1 does not occur spontaneously, whereas reaction 2 is exergonic and occurs in all cells. It is important to note that reactions 1 and 2 share a common intermediate; P_i (inorganic phosphate, HPO_4^{2-}), and these two reactions can be coupled to give a new reaction (reaction 3):

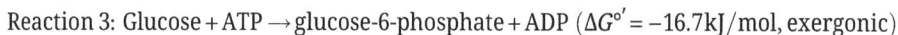

Reaction 3: Glucose + ATP → glucose-6-phosphate + ADP ($\Delta G^{\circ\prime}$ = −16.7kJ/mol, exergonic)

The energy released in reaction 2 is more than the energy required in reaction 1, so the free energy change for reaction 3, $\Delta G^{\circ\prime}$ is negative, and hence, synthesis of glucose-6-phosphate can occur when it is coupled with the breakdown of ATP into ADP and P_i. Thus, coupling of exergonic and endergonic reactions through a common intermediate is the key to the energy exchange in living systems. So, it is the breakdown of ATP that derives many endergonic processes in the cell; therefore, ATP is known as the energy currency of living systems.

As shown in Figure 4.4, ATP can be regenerated on phosphorylation of ADP, and the energy to phosphorylate ADP is provided by catabolic reactions occurring in the cell. The ATP cycle is a revolving door through which energy passes during its transfer from catabolic to anabolic pathways. Thus, ATP is continuously consumed and regenerated, and this process occurs at a very rapid rate.

Now, what are the factors responsible for high phosphoryl transfer potential of ATP?

Since $\Delta G^{\circ\prime}$ depends on the difference in free energies of the reactants and the products, to answer this question, one must consider the structures of both the ATP and its hydrolysis products, ADP and P_i. The four factors accounted for this are: (i)

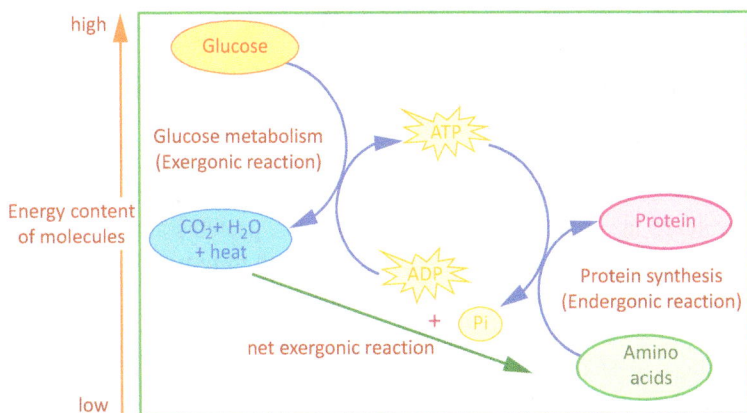

Figure 4.4: Schematic representation of coupled reactions in biological processes.

resonance stabilization, (ii) electrostatic repulsion, (iii) increase in entropy, and (iv) stabilization due to hydration.

(i) Resonance stabilization: The products ADP and P_i have more resonating structures than does ATP. Orthophosphate has a number of resonance structures of similar energy, whereas the γ-phosphoryl group of ATP has a smaller number.

(ii) Electrostatic repulsion: At physiological pH, the triphosphate unit of ATP carries four negative charges, and as they are in close proximity, they repel one another. These electrostatic repulsions will get reduced when ATP is hydrolyzed.

(iii) Increase in entropy: The entropy of products obtained on the hydrolysis of ATP is greater as there are two molecules instead of one single ATP molecule.

(iv) Hydration: Water molecules can bind more effectively to ADP and P_i, thus further stabilizing the products by hydration.

ATP must be generated in sufficient amounts to drive thermodynamically unfavorable biochemical reactions. The energy required to synthesize ATP is provided by the breakdown of other chemicals. The energy released in the cleavage of bonds is used for the coupling reaction of ADP and P_i to produce ATP. The γ-phosphorus of ATP can be transferred to a molecule which has lower energy of hydrolysis than ATP, for example, glucose 6-phosphate but the reverse reaction is not possible without putting extra energy. Thus, the ATP hydrolyzed in the cell could only be regenerated from phosphate compounds having the same or higher energy of hydrolysis than ATP itself such as creatine phosphate.

4.5 Carbohydrates and Energy Metabolism

The overall process of energy generation from the oxidation of foodstuffs in higher organisms takes place in three stages as shown in Figure 4.5. The food we consume mainly contains polysaccharides (carbohydrates), proteins, and fats. In the first stage, these molecules broke down into smaller units in the process of digestion. Polysaccharides are hydrolyzed into simple sugars such as glucose, proteins are hydrolyzed into 20 different amino acids, and lipids (fats) are hydrolyzed to fatty acids and glycerol. This stage is referred to as digestion, which occurs in the mouth, stomach, and small intestine, and no useful energy is captured in this stage.

In the second stage, these monomers are degraded to some simple units that play a crucial role in metabolism. Simple sugars, several amino acids, fatty acids, and glycerol are converted into the acetyl unit of acetyl CoA. Some amount of ATP is generated at this stage.

The third stage consists of CAC (citric acid cycle) and oxidative phosphorylation, in which acetyl group in the acetyl CoA is completely oxidized to carbon dioxide and produces ATP. For each acetyl group that is oxidized, total four pairs of electrons are transferred (three to NAD^+ and one to FAD) and then a proton gradient is created as

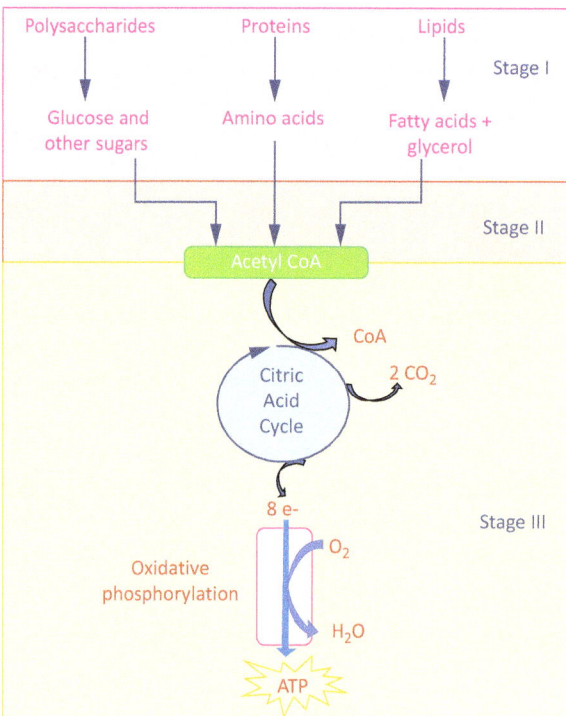

Figure 4.5: Different stages of catabolism.

Table 4.1: Types of chemical reactions in metabolism.

Reaction	Function	Example
Oxidation–reduction	Transfer of electron	1) Succinate + FAD $\rightleftharpoons$ fumarate + FADH$_2$ (citric acid cycle) 2) Malate + NAD$^+$ $\rightleftharpoons$ oxaloacetate + NADH + H$^+$ (citric acid cycle)
Hydrolytic	Bonds – cleavage with addition of water	Dipeptide + H$_2$O $\rightleftharpoons$ amino acid 1 + amino acid 2 (hydrolysis of proteins)
Isomerization	Rearrangements of atoms to generate isomers	Citrate $\rightleftharpoons$ isocitrate (citric acid cycle)
Group transfer	Transfer of a functional group from one molecule to another	Glucose + ATP $\rightleftharpoons$ glucose – 6 – phosphate + ADP (glycolysis)
Ligation involving ATP cleavage	Formation of covalent bonds	Pyruvate + CO$_2$ + ATP + H$_2$O $\rightleftharpoons$ oxaloacetate + ADP + P$_i$ + H$^+$ (formation of oxaloacetate)
Carbon bond cleavage (other than oxidation and hydrolysis)	Two substrates generating one product or vice versa. A double bond is formed when products are CO$_2$ and H$_2$O	Fructose 1, 6 – bisphosphate $\rightleftharpoons$ DHAP + GAP(glycolysis) DHAP (dihydroxyacetone phosphate) and GAP (glyceraldehyde 3-phosphate)

electrons move from these electron carriers to oxygen, and resulted in the generation of ATP.

The thousands of reactions that occur in metabolic pathways can be divided into six main categories (given in Table 4.1), and specific reactions of each type can appear repeatedly in a particular metabolic pathway. All these six reactions can proceed in either direction, depending on several factors such as the standard free energy of that reaction, the concentration of reactants, and products inside the cell.

4.5.1 Glycolysis: An Energy Conversion Pathway in Many Organisms

The first metabolic pathway which we discuss is glycolysis, and it is defined as the linked series of reactions that converts one molecule of glucose to two molecules of pyruvate with concomitant production of two molecules of ATP. In eukaryotic cells, glycolysis occurs in the cytoplasm and is anaerobic in nature (i.e., it does not require oxygen). The product of this metabolic pathway, pyruvate, can then be converted under anaerobic conditions, to either ethanol (alcoholic fermentation) or lactate (lactic acid fermentation). Under aerobic conditions (in the presence of oxygen), pyruvate can be completely oxidized to carbon dioxide (CO_2), producing enormous amount of ATP (Figure 4.6).

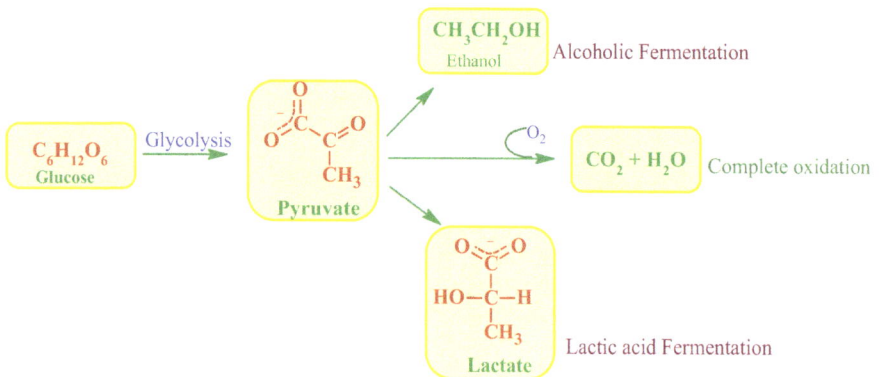

Figure 4.6: Metabolism of glucose.

The glycolytic pathway consists of two phases: (i) preparatory phase and (ii) the payoff phase (Figure 4.7), which consists of the following steps:

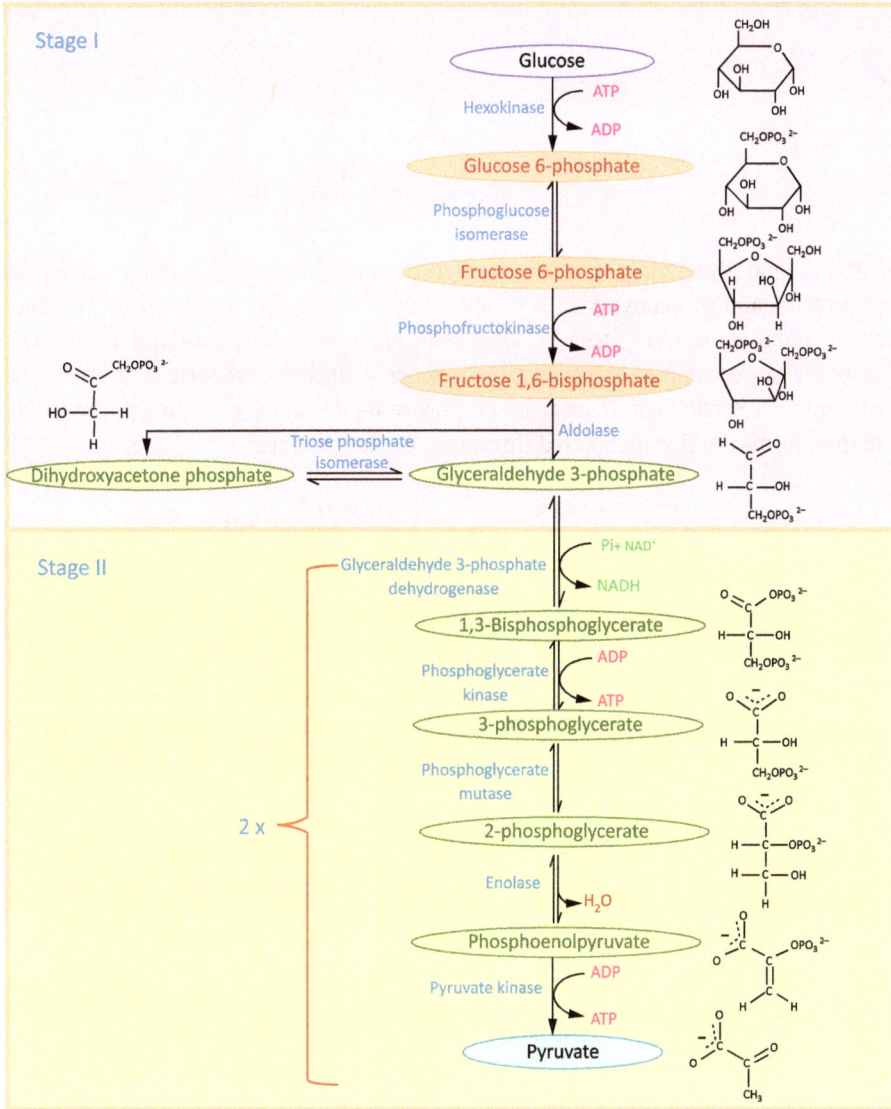

Figure 4.7: Stages of glycolysis.

(i) Glucose enters the cells through specific proteins and is phosphorylated by ATP to produce glucose 6-phosphate. This reaction is catalyzed by the enzyme *hexokinase* that requires Mg^{2+} ion for its activity. Magnesium forms a complex with ATP:

Glucose   Glucose-6-phosphate (G-6P)

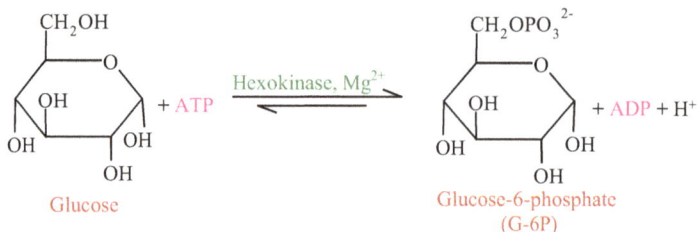

(ii) The second glycolytic step is the isomerization of glucose 6-phosphate to fructose 6-phosphate, and this conversion is catalyzed by *phosphoglucose isomerase*. This conversion involves several steps because both glucose 6-phosphate and fructose 6-phosphate are present in the cyclic form. *Phosphoglucose isomerase* first opens the six-membered cyclic form (pyranose) of glucose 6-phosphate, catalyzes the reaction, and then forms the five-membered (furanose) cyclic structure:

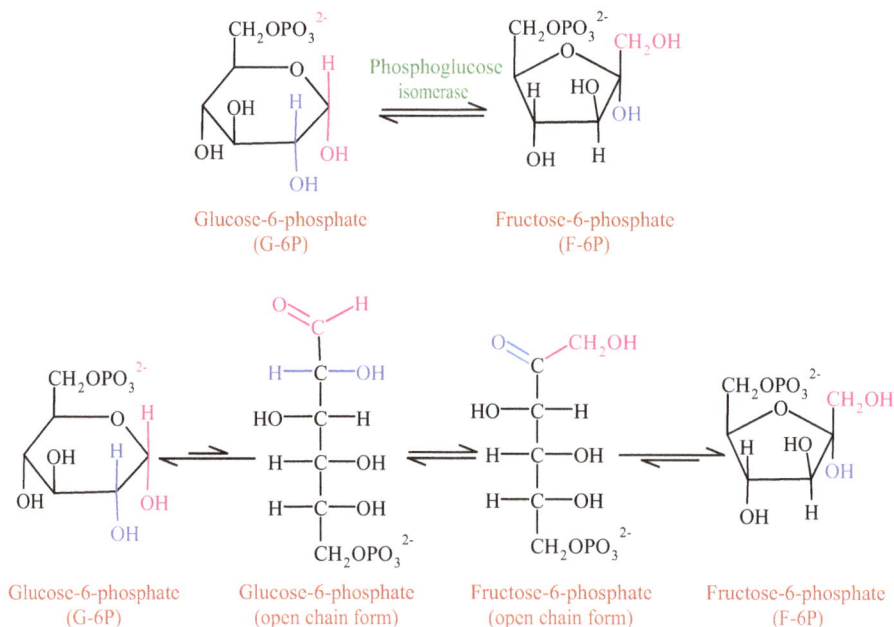

Glucose-6-phosphate (G-6P)   Fructose-6-phosphate (F-6P)

Glucose-6-phosphate (G-6P)   Glucose-6-phosphate (open chain form)   Fructose-6-phosphate (open chain form)   Fructose-6-phosphate (F-6P)

(iii) The isomerization step is followed by the second phosphorylation reaction. Fructose 6-phosphate is phosphorylated to yield fructose 1,6-bisphosphate, and this step is catalyzed by *phosphofructokinase*. The prefix bis- in bisphosphate indicates that the two separate monophosphoryl groups are present.

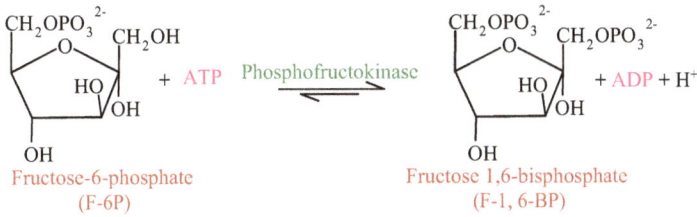

Fructose-6-phosphate
(F-6P)

+ ATP   Phosphofructokinase   + ADP + H⁺

Fructose 1,6-bisphosphate
(F-1, 6-BP)

(iv) In the next step, fructose 1,6-bisphosphate is cleaved to yield two different triose phosphates, glyceraldehyde 3-phosphate (GAP, an aldose) and dihydroxyacetone phosphate (DHAP, a ketose), completing stage 1 of glycolysis. This step of the glycolytic pathway is reversible and is catalyzed by the enzyme *aldolase*:

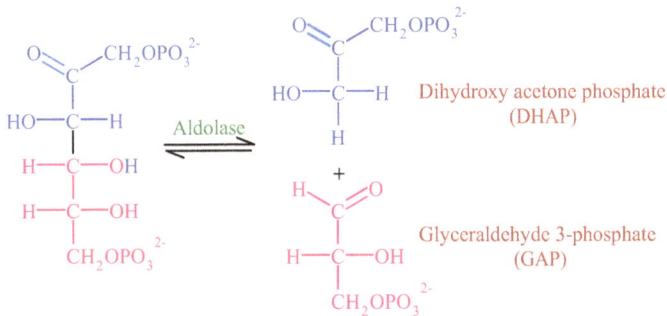

Aldolase

Dihydroxy acetone phosphate
(DHAP)

Glyceraldehyde 3-phosphate
(GAP)

(v) GAP is further utilized in glycolysis, whereas DHAP is not. DHAP is quickly and reversibly converted to GAP. *Triose phosphate isomerase* catalyzed this isomerization step. The hexose undergoes phosphorylation at C-1 and C-6, and cleaved to form two molecules of GAP. This is the last step of preparatory phase of glycolysis:

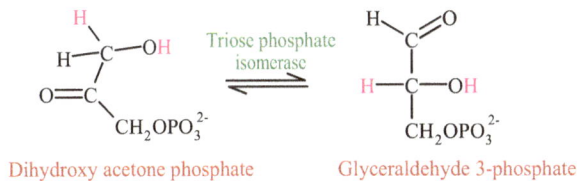

Triose phosphate
isomerase

Dihydroxy acetone phosphate   Glyceraldehyde 3-phosphate

(vi) The preceding steps have utilized two molecules of ATP, and converted one molecule of glucose into two molecules of GAP. After the preparatory phase, the first step in the pay-off phase is the oxidation of GAP to 1,3-bisphosphoglycerate (1,3-BPG), and this reaction is catalyzed by *glyceraldehyde 3-phosphate dehydrogenase*:

Glyceraldehyde 3-phosphate + NAD$^+$ + Pi $\underset{\text{dehydrogenase}}{\overset{\text{Glyceraldehyde 3-phosphate}}{\rightleftharpoons}}$ 1,3-Bisphosphoglycerate (1,3-BPG) + NADH + H$^+$

Glyceraldehyde 3-phosphate

1,3-Bisphosphoglycerate (1,3-BPG)

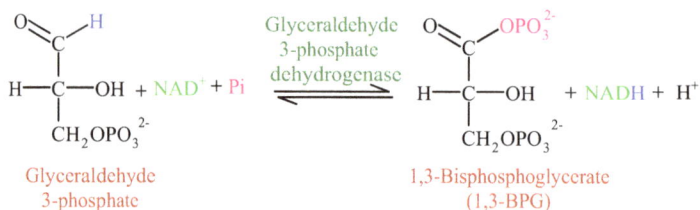

This reaction is a combination of two steps: (i) in the first step, an aldehyde (GAP) is oxidized to carboxylic acid by NAD$^+$; and (ii) the carboxylic acid and P$_i$ joined to form the acyl-phosphate:

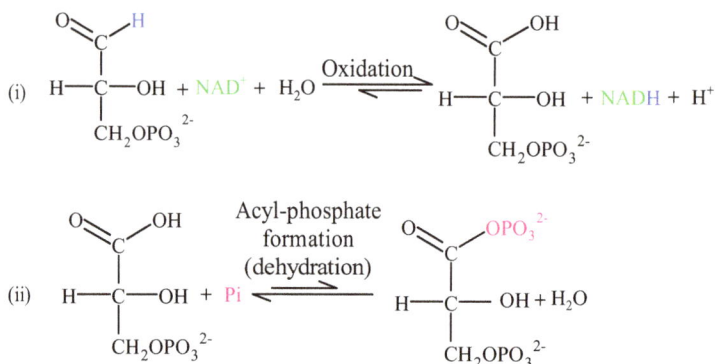

(i) aldehyde + NAD$^+$ + H$_2$O $\overset{\text{Oxidation}}{\rightleftharpoons}$ carboxylic acid + NADH + H$^+$

(ii) carboxylic acid + Pi $\underset{}{\overset{\text{Acyl-phosphate formation (dehydration)}}{\rightleftharpoons}}$ acyl-phosphate + H$_2$O

(vii) 1,3-BPG is an energy-rich molecule and has high phosphoryl transfer potential than ATP. Thus, this step can be employed to synthesize ATP from ADP. *Phosphoglycerate kinase* catalyzes the transfer of phosphoryl group from the carboxyl group of 1,3-BPG to ADP, which results in the formation of 3-phosphoglycerate and ATP:

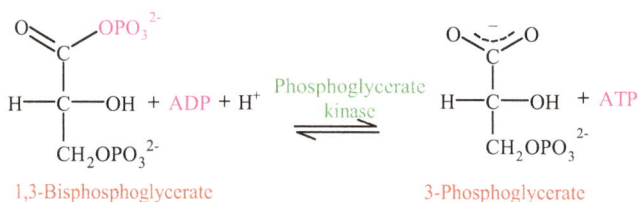

1,3-Bisphosphoglycerate + ADP + H$^+$ $\overset{\text{Phosphoglycerate kinase}}{\rightleftharpoons}$ 3-Phosphoglycerate + ATP

(viii) The remaining steps of glycolysis involve the conversion of 3-phosphoglycerate into pyruvate, resulting in the generation of second molecule of ATP from ADP. The first step, that is, the rearrangement step, is catalyzed by the enzyme *phosphoglycerate mutase*, in which shift of the phosphoryl group from C-3 to C-2 of glycerate takes place. A *mutase* is an enzyme that catalyzes the shifting of a functional group, such as phosphoryl group, within the same molecule:

3-Phosphoglycerate Phosphoglycerate mutase 2-Phosphoglycerate

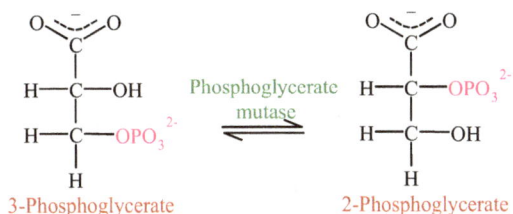

(ix) The second last step is the reversible removal of water molecule from 2-phosphoglycerate to yield phosphoenolpyruvate (PEP), and this step is catalyzed by *enolase*:

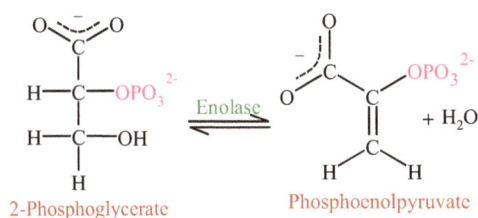

2-Phosphoglycerate Enolase $+ H_2O$ Phosphoenolpyruvate

(x) The last step of glycolysis is the transfer of phosphoryl group from PEP to ADP, and then PEP gets converted into the more stable pyruvate. The irreversible step is catalyzed by *pyruvate kinase*:

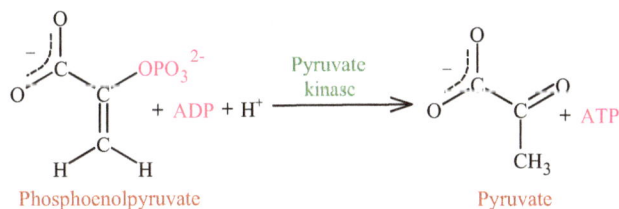

Phosphoenolpyruvate $+ ADP + H^+$ Pyruvate kinase $+ ATP$ Pyruvate

The net reaction of glycolysis is

$$\text{Glucose} + 2\,P_i + 2\,\text{ADP} + 2\,\text{NAD}^+ \rightarrow 2\,\text{pyruvate} + 2\,\text{ATP} + 2\,\text{NADH} + 2\,\text{H}^+ + 2\,\text{H}_2\text{O}$$

Thus, two molecules of ATP are generated in the conversion of one molecule of glucose into two molecules of pyruvate.

4.5.2 Alcoholic and Lactic Acid Fermentation

The sequence of reactions for the transformation of glucose into pyruvate is same in most types of cells and in many organisms. However, the fate of pyruvate is different. Pyruvate can be either converted into alcohol, lactate, or carbon dioxide. The first two reactions, that is, the conversion of pyruvate into alcohol and lactate, take place in the absence of oxygen, and are known as fermentation reactions. Under aerobic conditions, the most common reaction that occurs in multicellular organisms and in some unicellular organisms is the conversion of pyruvate into carbon dioxide and water, via CAC and the electron transport chain. In these processes, oxygen acts as the final electron acceptor. We now discuss the possible fates of pyruvate under different conditions (Figure 4.8).

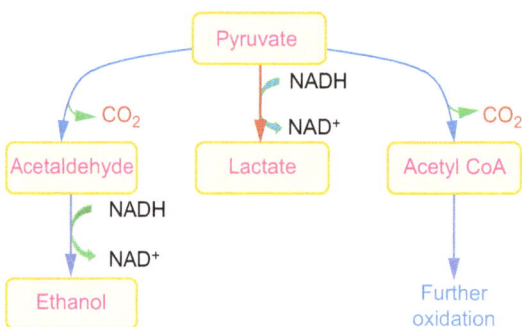

Figure 4.8: Diverse fate of pyruvate.

Glycolysis provides only two molecules of ATP but sometimes some amount of energy is better than no energy. So, some organisms undergo fermentation process in anaerobic condition to generate ATP by performing glycolysis and one extra step. Alcoholic fermentation takes place in yeast cells which leads to the production of beer, bread, and wine, whereas lactic acid fermentation occurs in our muscle cells when we run or do some strenuous exercise.

4.5.2.1 Alcoholic Fermentation

Alcoholic fermentation takes place in yeast and several other microorganisms. This reaction occurs in two steps: decarboxylation of pyruvate takes place in the first step, which is catalyzed by *pyruvate decarboxylase*. This enzyme requires the coenzyme thiamine pyrophosphate for its activity. The second step is the reduction of aldehyde to ethanol by NADH, and this reaction is catalyzed by *alcohol dehydrogenase*. This reaction generates NAD^+:

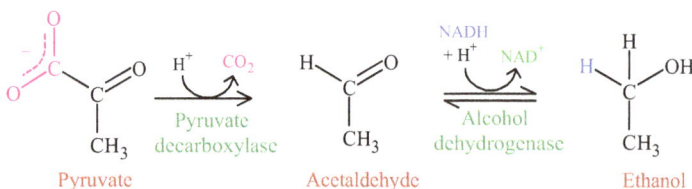

Alcoholic fermentation is used for commercial production of beer, wine, bread, and cheese. In bread making, the bread grows or rises due to the presence of carbon dioxide in between the wheat protein that is known as gluten. The specific smell bread gets owing to the presence of ethanol. In beverages (beer and wine), CO_2 is responsible for the bubbly appearance of the liquid. The net reaction involved in the conversion of glucose to ethanol is

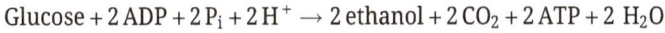

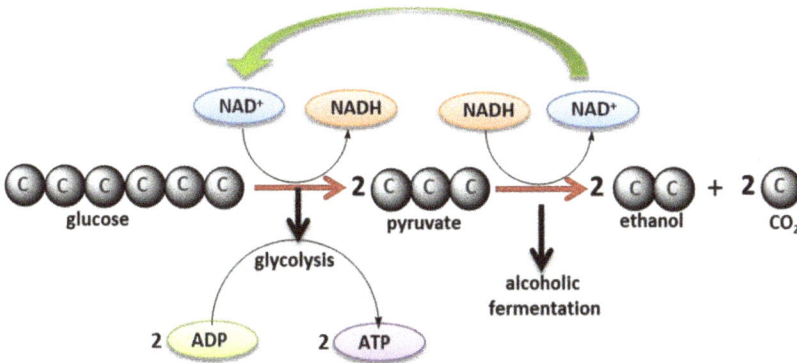

$$\text{Glucose} + 2\,\text{ADP} + 2\,\text{P}_i + 2\,\text{H}^+ \rightarrow 2\,\text{ethanol} + 2\,\text{CO}_2 + 2\,\text{ATP} + 2\,\text{H}_2\text{O}$$

4.5.2.2 Lactic Acid Fermentation

Under normal conditions, the muscle cells in the presence of oxygen carry out normal cellular respiration while in the absence or lack of oxygen it will undergo lactic acid fermentation. Lactate is formed from pyruvate in many microorganisms, and this process is known as lactic acid fermentation. The reduction of pyruvate to lactate by NADH is catalyzed by *lactate dehydrogenase*:

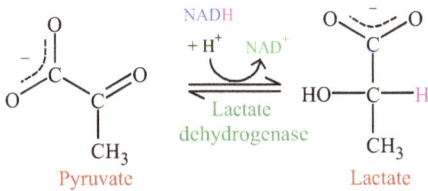

Lactic acid makes the muscle sore and stiff after strenuous physical exercise. These muscle fibers cannot get rid of lactic acid by any mechanism, so they have to wait for the acid to be substantially washed away by blood stream into the liver, and then the liver eliminates lactic acid from the system. The overall reaction involved in the conversion of glucose into lactate is

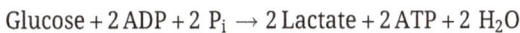

$$\text{Glucose} + 2\,\text{ADP} + 2\,\text{P}_i \rightarrow 2\,\text{Lactate} + 2\,\text{ATP} + 2\,\text{H}_2\text{O}$$

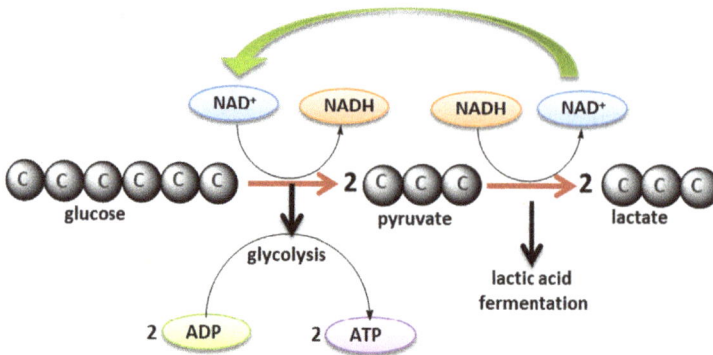

In alcoholic as well as lactic acid fermentation, although NAD^+ and NADH do not appear in the final equation, they are important for the overall process. The NADH generated in the oxidation of GAP is taken up for the reduction of acetaldehyde to alcohol in alcoholic fermentation, and to lactate in lactic acid fermentation. NAD^+ is regenerated in the reduction of pyruvate to ethanol or lactate, and sustains the glycolytic pathway under anaerobic conditions.

Only a small amount of energy is released in its anaerobic conversion into ethanol and lactate. Most of the energy can be extracted by the aerobic processing, that is, when pyruvate generated from glucose undergoes oxidative decarboxylation to produce acetyl CoA, this step is catalyzed by enzyme *pyruvate dehydrogenase*. This oxidation process occurs in a series of reactions known as CAC, also known as tricarboxylic acid cycle or Kreb's cycle. Most fuel molecules such as carbohydrates, amino acids, and fatty acids enter the CAC as acetyl CoA. In eukaryotes, glycolysis occurs in the cytoplasm, whereas CAC takes place in matrix of the mitochondria.

4.5.3 Citric Acid Cycle

The major function of CAC is to act as the final common pathway for the oxidation of carbohydrate, lipids, and proteins. This is because fatty acids, glucose, and many amino acids are all metabolized to acetyl CoA or intermediates of the cycle. The cycle also acts as the precursor for the building blocks of many important biological molecules such as nucleotide bases, amino acids, and porphyrin. Oxaloacetate, one of the CAC components, is also an important precursor to glucose. The enzymes that facilitate the CAC are present in the mitochondrial matrix, either in free form or anchored to the inner surface of the inner mitochondrial membrane. The overview of CAC is shown in Figure 4.9.

How does CAC transform fuel molecules into ATP?

The CAC involves a series of oxidation–reduction reactions that oxidize an acetyl group into two molecules of carbon dioxide. The oxidation process generates high-energy electrons that are used to synthesize ATP. It is important to note here that CAC

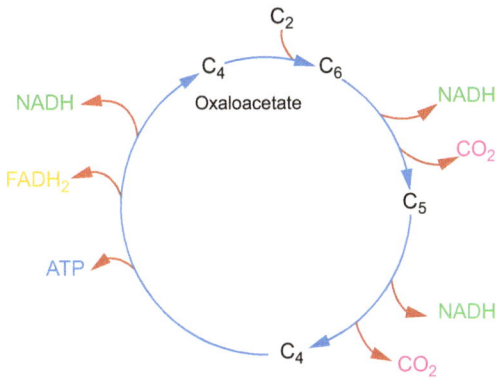

Figure 4.9: Overview of the citric acid cycle.

neither uses oxygen as a reactant nor generate a large amount of ATP but it only harvests high-energy electrons from carbon compounds (fuel molecules), and use these electrons to reduce NAD^+ and FAD to NADH and $FADH_2$, respectively. Electrons released in this process then move via a series of membrane proteins (known as electron transport chain), which then create potential gradient across the inner mitochondrial membrane. These protons then move through ATP synthases to generate ATP from ADP and inorganic phosphate. These electron carriers resulted in the generation of nine molecules of ATP when they are oxidized by oxygen molecules in oxidative phosphorylation.

The pyruvate dehydrogenase generates a link between glycolysis and CAC. Acetyl CoA is generated from the oxidative decarboxylation of pyruvate in the mitochondrial matrix by the enzyme *pyruvate dehydrogenase* complex, which works as a fuel for CAC:

$$Pyruvate + CoA + NAD^+ \rightarrow acetyl\,CoA + CO_2 + NADH + H^+$$

The CAC involves the following steps:

Step 1: Citrate synthase forms citrate from acetyl coenzyme A and oxaloacetate
The first reaction is the condensation of acetyl group (two-carbon unit) of acetyl CoA with oxaloacetate (four-carbon unit), and this step, aldol condensation, followed by hydrolysis is catalyzed by the enzyme *citrate synthase*:

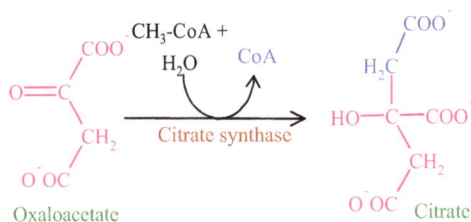

Oxaloacetate reacts with acetyl CoA to form citryl CoA, and then hydrolysis of citryl CoA results in the formation of citrate and CoA:

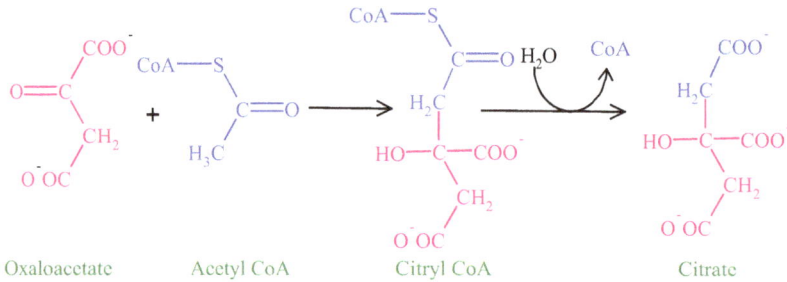

Oxaloacetate   Acetyl CoA   Citryl CoA   Citrate

Step 2: Isomerization of citrate to isocitrate

In this step, citrate is isomerized into isocitrate with the help of enzyme *aconitase* (*aconitate hydratase*), which contains iron in the Fe^{2+} state in the form of an iron–sulfur protein (Fe:S):

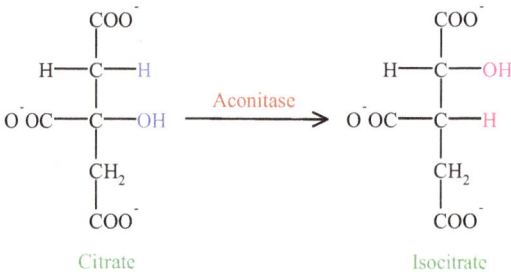

Citrate   Isocitrate

This isomerization step is accomplished in two steps: dehydration step followed by hydration resulted in an exchange of H and OH:

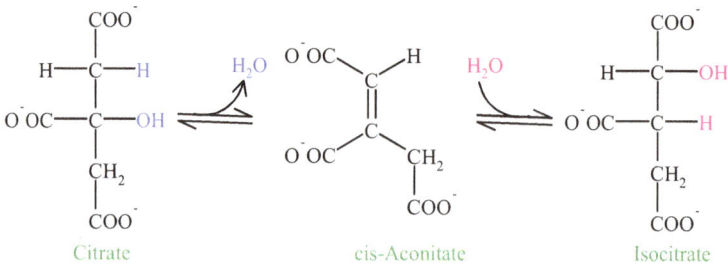

Citrate   cis-Aconitate   Isocitrate

Step 3: Isocitrate is oxidized and decarboxylated to alpha-ketoglutarate

This is the first of four oxidation–reduction reactions that occur in CAC. Isocitrate is oxidatively decarboxylated to oxalosuccinate, and this step is catalyzed by *isocitrate dehydrogenase*, While bound to the enzyme, oxalosuccinate, an unstable β-ketoacid, is decarboxylated to form α-ketoglutarate. This oxidation step produces the first poten-tial electron carrier, NADH, in the cycle:

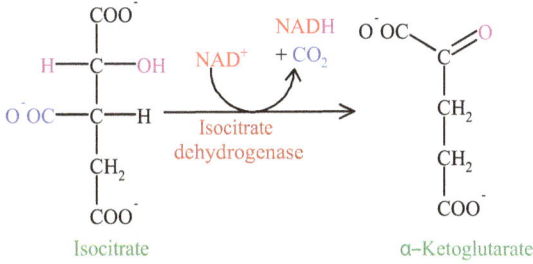

Isocitrate → α−Ketoglutarate

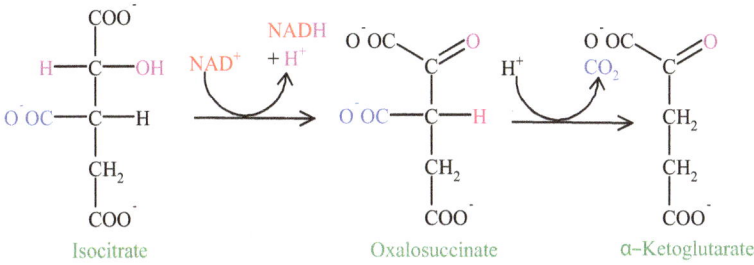

Isocitrate → Oxalosuccinate → α−Ketoglutarate

Step 4: α-Ketoglutarate undergoes oxidative decarboxylation to form succinyl coenzyme A

In this step, succinyl CoA is formed by the oxidative decarboxylation of α-ketoglutarate, and this reaction is catalyzed by the *a-ketoglutarate dehydrogenase complex*. This step closely resembles the oxidative decarboxylation of pyruvate:

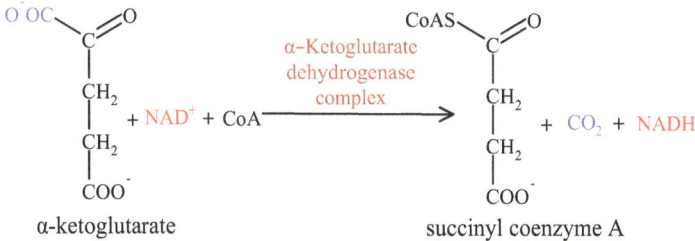

α-ketoglutarate → succinyl coenzyme A

Step 5: ATP is generated from succinyl CoA

This step involves the conversion of succinyl CoA, which is an energy-rich thioester compound to succinate, and is catalyzed by *succinyl CoA synthetase (succinate thioki-nase)*. This is the only step in CAC that directly generates a high-energy phosphoryl transfer compound:

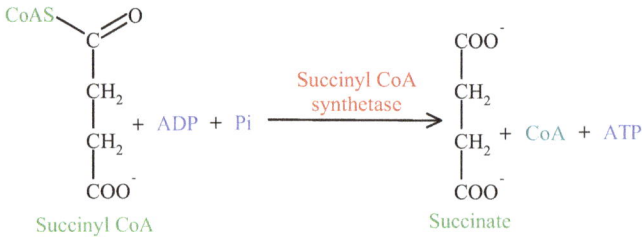

Succinyl CoA → Succinate (Succinyl CoA synthetase) + ADP + Pi → + CoA + ATP

Step 6: Oxidation of succinate to fumarate

Succinate is oxidized to fumarate by dehydrogenation reaction in the presence of enzyme *succinate dehydrogenase*. In this step, FAD is the hydrogen acceptor rather than NAD^+ as the free energy is not sufficient to reduce NAD^+:

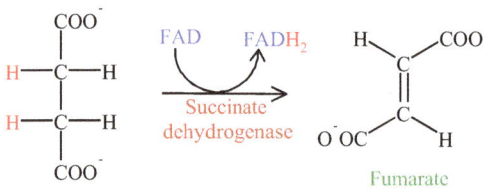

Step 7: Hydration of fumarate to form malate

Fumarase (fumarate hydratase) catalyzes a stereospecific trans addition of H^+ and OH^-. In this step, addition of the hydroxyl group to only one side of the double bond of fumarate takes place, resulting in the formation of only L-isomer:

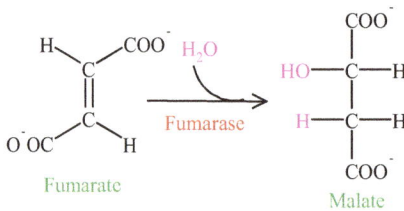

Step 8: Oxidation of malate to oxaloacetate

In the last step of the CAC, malate is oxidized to oxaloacetate. This step requires NAD^+, and is catalyzed by enzyme *malate dehydrogenase*:

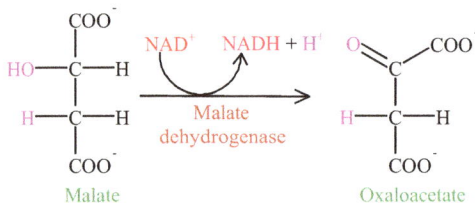

The net reaction of CAC is

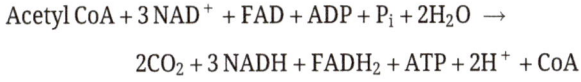

$$\text{Acetyl CoA} + 3\,\text{NAD}^+ + \text{FAD} + \text{ADP} + P_i + 2H_2O \ \rightarrow$$

$$2CO_2 + 3\,\text{NADH} + \text{FADH}_2 + \text{ATP} + 2H^+ + \text{CoA}$$

and a cyclic representation is shown in Figure 4.10.

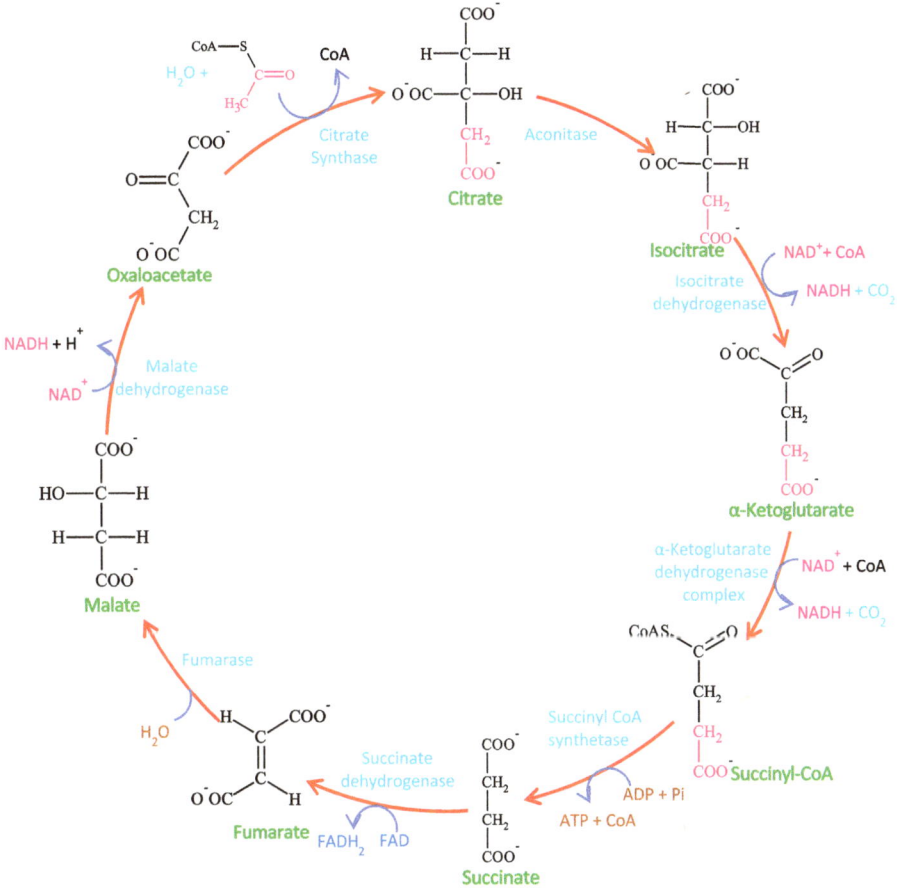

Figure 4.10: Citric acid cycle.

References

[1]   Berg JM, Tymoczko JL, Gatto Jr GJ. Lubert Stryer. Biochemistry. Freeman Macmillan. 8th Edition. ISBN-13. 2006:978–1.

[2]   Hames D, Hooper N. Instant notes biochemistry. Taylor & Francis; 2006 Oxfordshire, United Kingdom. Sep 27

[3]   http://samples.jbpub.com/9780763776633/76633_ch07_5589.pdf

[4]   Lehninger AL, Nelson DL, Cox MM. Lehninger Principles of Biochemistry. Fourth ed. New York: W.H. Freeman; 2005.

[5]   Palmer M. Human metabolism: Lecture notes.

Chapter 5
Nucleic Acids

The versatility and elegant simplicity of DNA (deoxyribonucleic acid) double helix dictate life. Although, nucleic acid was first identified as 'nuclein' by Friedrich Miescher, a Swiss physician and biologist in 1869, yet it took more than 70 years to demonstrate that it is the molecule that carries genetic information. DNA of a single cell contains all genetic information necessary for processes of life. The double-helical structure of DNA was discovered by James D. Watson and Francis H. C. Crick in 1953, using the X-ray diffraction data of Rosalind Franklin, and this marvelous discovery proved to be the significant turning point that paved the way to the development of biomedical science and modern biology. Structurally, DNA is a flexible molecule which can adopt numerous unusual structures depending on the solution conditions. To understand the structure of B-form of DNA (natural/native form), it is important to understand the individual components of DNA. Nucleic acids (DNA and RNA (ribonucleic acid)) are long linear polymers which are made up of monomer units called nucleotides. All nucleotides are made up of three components: a nitrogen heterocyclic base, a pentose sugar, and a phosphate group. They are termed as nucleic acids because they are found in nucleus and comprise phosphoric acid (phosphate) as one of its components. The two types of nucleic acids that occur in cells are DNA and RNA. DNA acts as a carrier of genetic information and is found in the nucleus of the cell, whereas RNA is present in both the nucleus and cytoplasm.

5.1 Components of Nucleic Acids

Both DNA and RNA are biopolymers, which are made up of monomer units called nucleotide. Nucleotide is the building block of nucleic acids which consists of three components:
(i)   pentose sugar
(ii) nitrogenous base
(iii) inorganic phosphate

5.1.1 Pentose Sugar

It is of two types:
(i)   Deoxyribose
(ii) Ribose

In DNA, the pentose sugar is 2′-deoxyribose, whereas in RNA, it is ribose where hydrogen at the 2′-position is replaced by the hydroxyl group. Why primes? The carbon atoms

https://doi.org/10.1515/9783110793765-005

in sugars are numbered as 1', 2', and so on, to differentiate them from atoms of nitrogenous bases. The chemical structures of these two sugars are shown in Figure 5.1.

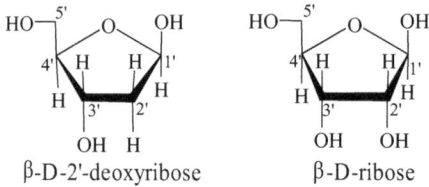

Figure 5.1: Structure of sugars.

5.1.2 Nitrogenous Bases

Two types of nitrogenous bases present in nucleic acids are monocyclic pyrimidines and bicyclic purines. The two purine bases found in DNA and RNA are adenine (A) and guanine (G). Adenine has an amino group (-NH$_2$) at the C-6 position of purine ring and guanine has an amino group (-NH$_2$) at C-2 position and carbonyl group (-C = O) at C-6 position. The pyrimidine bases present in DNA are thymine (T) and cytosine (C), whereas uracil (U) is found in RNA, in place of thymine. Thymine contains a methyl group (-CH$_3$) at the C-5 position with carbonyl group (-C = O) at C-2 and C-4 positions. Cytosine contains a carbonyl group at C-2 with an amino group at C-4, whereas uracil contains carbonyl group (-C = O) at C-2 and C-4 positions (Figure 5.2).

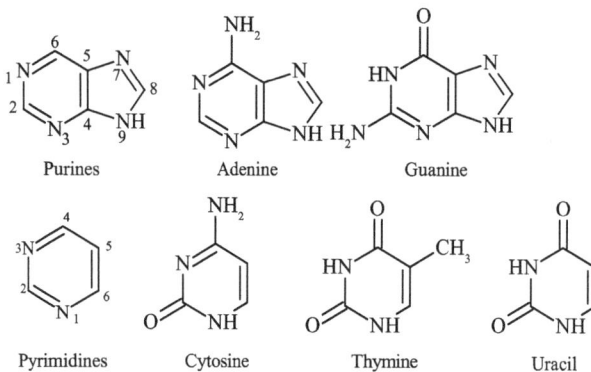

Figure 5.2: Structure of nitrogenous bases: purines (adenine and guanine) and pyrimidines (cytosine, thymine, and uracil).

5.1.3 Inorganic Phosphate

Phosphate group in nucleic acids is derived from phosphoric acid, and is attached through the oxygen atom of hydroxyl group of the 5'-carbon of the pentose sugar. At physiological pH, inorganic phosphate exists primarily as a nearly equal mixture of dihydrogen phosphate and monohydrogen phosphate. Thus, phosphate solution functions as effective buffer at pH 7.4.

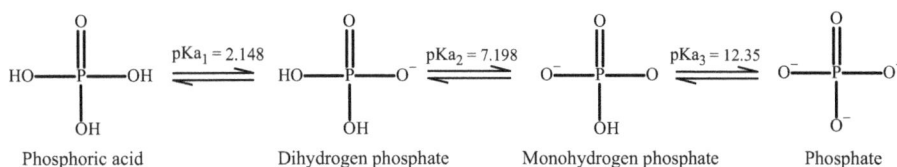

| Phosphoric acid | Dihydrogen phosphate | Monohydrogen phosphate | Phosphate |

$pKa_1 = 2.148$ $pKa_2 = 7.198$ $pKa_3 = 12.35$

5.2 Nucleosides and Nucleotides

Each nitrogenous base combines with a molecule of sugar (deoxyribose or ribose) to form a nucleoside. It is N-9 in purine and N-1 in pyrimidines which are attached to the C-1' of sugars. The structures of DNA nucleosides are shown in Figure 5.3. If the sugar residue is deoxyribose, then it is called deoxyribonucleoside, whereas in case of a ribose sugar, nucleoside is termed as ribonucleoside. The bond between the sugar and the base is called the N-glycosidic bond. The base is free to rotate around the glycosidic bond, and the two standard conformations of the base around glycosidic bond are *syn* and *anti*.

When phosphoric acid is esterified to one of the hydroxyl groups of sugar of the nucleoside, a nucleotide is formed. A single nucleotide unit consists of a purine or a pyrimidine base, deoxyribose or ribose sugar, and a phosphate group. The point of attachment of phosphate group can be 3'- or 5'- in the deoxyribose sugar. A nucleotide is named according to the number of phosphates (mono-, di-, or tri-) attached to the sugar, the type of sugar (deoxy- or ribose), and the nature of base. For example, adenosine (nucleoside) esterified with phosphoric acid to form adenosine monophosphate. The nomenclature of nucleic acids is given in Table 5.1.

The structures and names of four deoxyribonucleotides are shown in Figure 5.4. Methylated forms of bases are also present in DNA, and nucleotides with phosphate group at position 3' are also present in the cell.

Nucleoside diphosphates (NDPs) and nucleoside triphosphates (NTPs) are formed when one or two phosphate groups are added to the first phosphate group of a nucleoside molecule via pyrophosphate linkage. NTPs such as adenosine triphosphate (ATP) play an important role in many cellular processes, and hydrolysis of ATP releases a

Table 5.1: Nomenclature of nucleic acids.

	Nucleosides	
Base	DNA	RNA
Adenine	Deoxyadenosine	Adenosine
Thymine	Thymidine	–
Guanine	Deoxyguanosine	Guanosine
Cytosine	Deoxycytidine	Cytidine
Uracil	–	Uridine
	Nucleotides	
Adenine	Deoxyadenylate	Adenylate
Thymine	Thymidylate	–
Guanine	Deoxyguanylate	Guanylate
Cytosine	Deoxycytidylate	Cytidylate
Uracil	–	Uridylate

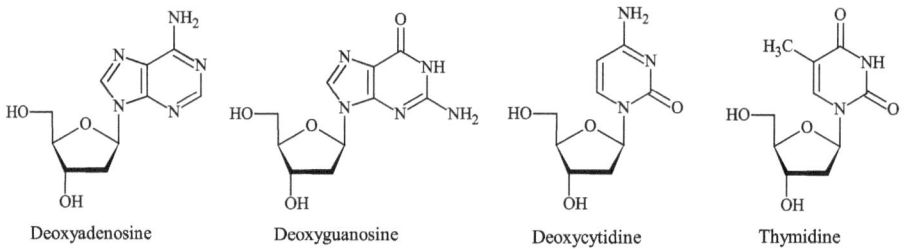

Figure 5.3: Structure of DNA nucleosides.

large amount of energy which is used to drive unfavorable reactions. The structures of NMP (nucleoside monophosphate), NDP, and NTP are shown in Figure 5.5.

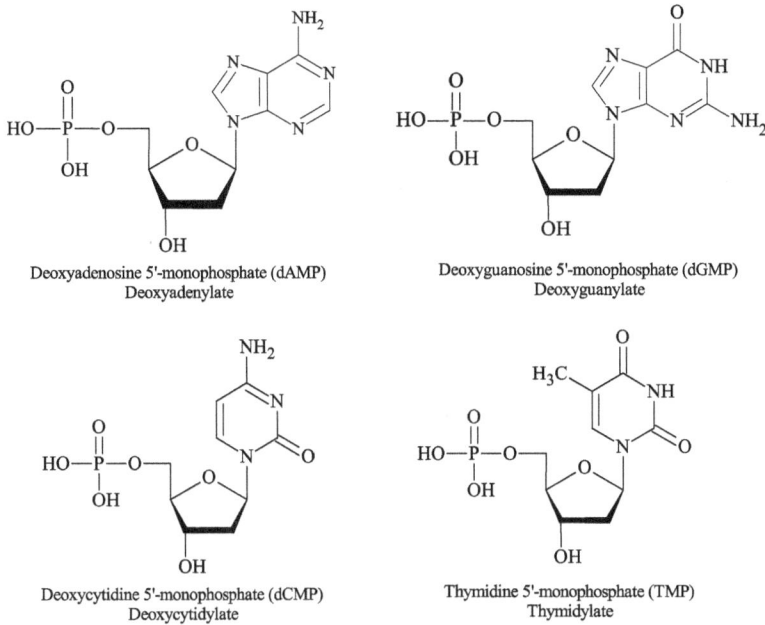

Deoxyadenosine 5'-monophosphate (dAMP)
Deoxyadenylate

Deoxyguanosine 5'-monophosphate (dGMP)
Deoxyguanylate

Deoxycytidine 5'-monophosphate (dCMP)
Deoxycytidylate

Thymidine 5'-monophosphate (TMP)
Thymidylate

Figure 5.4: Structure of DNA nucleotides.

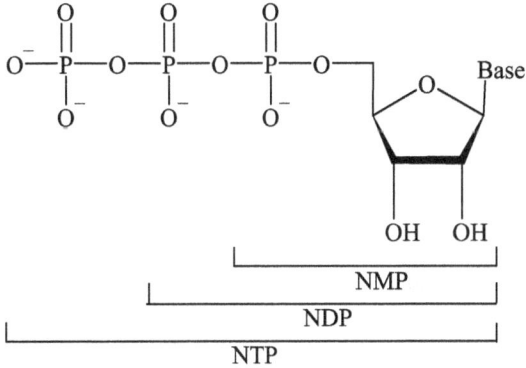

Figure 5.5: Structures of NMP, NDP, and NTP.

5.3 Synthesis of Nitrogenous Bases

5.3.1 Purines

The term "purine" was first coined by the German chemist Emil Fischer in 1884. The structure of a purine consists of a pyrimidine ring fused with an imidazole ring, and it exists in two tautomeric forms. Purine bases adenine, guanine, xanthine, hypoxanthine, theophylline, and theobromine are of great importance because of their occurrence in living tissues. Purine ring is also present in many coenzymes (ATP) and cofactors (NAD$^+$, FAD), which plays an important role in various biological processes. We have already discussed that adenine and guanine serve as an important constituent of nucleic acids; other purine bases also form an integral part of several plants. For example, theobromine is present in cacao plant; xanthine and hypoxanthine occur in tea extract and animal tissues; and theophylline forms an important constituent of tea leaves:

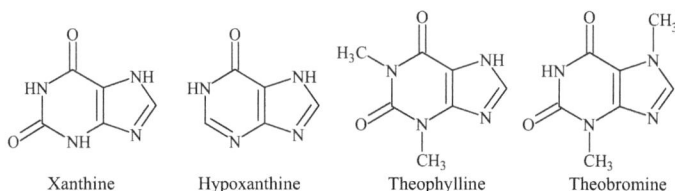

| Xanthine | Hypoxanthine | Theophylline | Theobromine |

The synthesis of purine for the first time was carried out by Fischer in 1898 using uric acid (1). Uric acid (1) on reaction with PCl$_5$ gives 2,6,8-trichloropurine (2), which was then converted to 2,6-diiodopurine using HI and PH$_4$I (3). The product was reduced using zinc dust to obtain purine (4). The trichloro compound (2) serves as an important intermediate in the synthesis of purine derivatives:

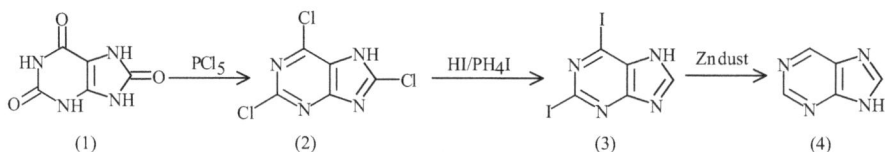

| (1) | (2) | (3) | (4) |

5.3.1.1 Adenine (6-Aminopurine)
Adenine occurs in tea extract and in pancreas of cattle. It plays a very important role in biochemical reactions including cellular respiration as it serves as the main component of energy-rich molecule ATP, NAD (nicotinamide adenine dinucleotide), FAD (flavin adenine dinucleotide), and coenzyme A. The various methods employed for the synthesis of adenine are discussed further.

i. Fischer's method (1897)
This method involves the synthesis of adenine from 2,6,8-trichloropurine. 2,6,8-trichloropurine on treatment with aqueous ammonia gives 6-amino-2,8-dichloropurine which on further reaction with HI yields adenine:

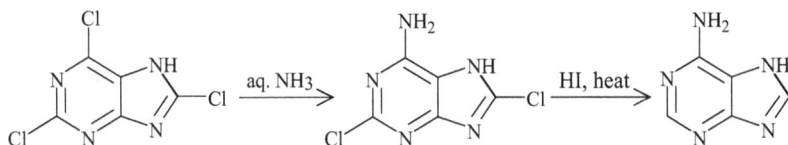

ii. Traube's method (1904)

Traube's synthesis is the most versatile method which is used to prepare any purine derivative. This method involves the synthesis of 4,6-diamino-2-mercaptopyrimidine using thiourea and propanedinitrile. The resulting intermediate is then nitrosated and reduced with ammonium bisulfide, and the obtained product is condensed with formic acid. The formyl derivative when heated in the presence of sodium salt at 250 °C gives 2-thiopurine which on treatment with H_2O_2 yields adenine:

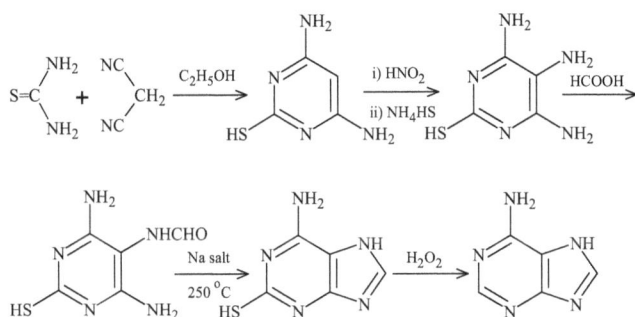

iii. Todd et al.'s method (1943)

This method consists of preparing adenine by condensing formamidine with phenylazomalononitrile. The resulting intermediate is then reduced to 4,5,6-triamino pyrimidine using H_2–Raney Ni which on hydrolysis of intermediate yields adenine:

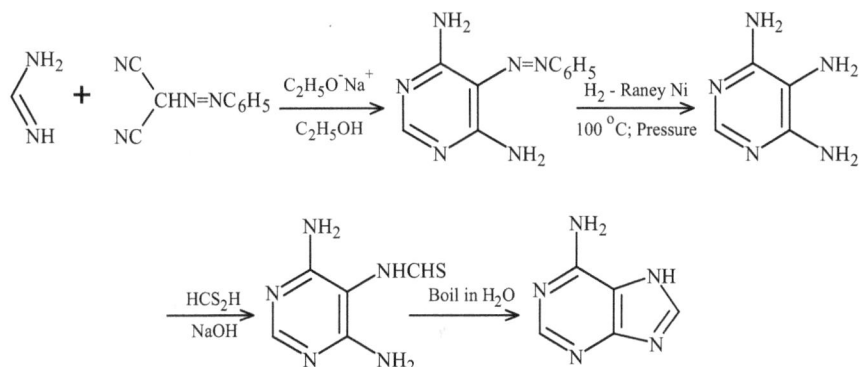

iv. Bredereck et al.'s method (1955)

This method is known as modified Traube's synthesis. This involves the treatment of 4,6-diamino-2-mercaptopyrimidine with methyl iodide. The intermediate was nitrosated and reduced to 4,5,6-triamino-2-methylthiopyrimidine, which on refluxing with formamide, and on reduction with Raney nickel gives adenine:

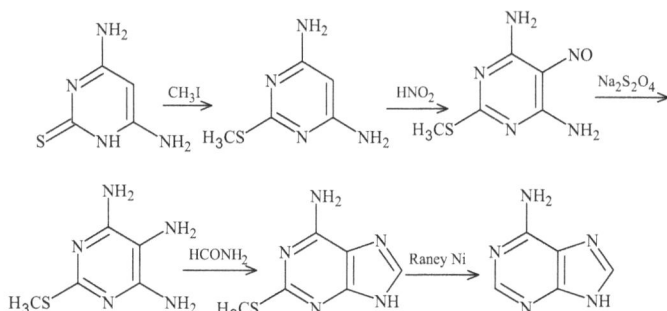

5.3.1.2 Guanine (2-Amino-6-Hydroxypurine)

Guanine exists in two tautomeric forms: keto form (major) and enol form (rare). It occurs in the pancreas of cattle, in certain fish scales, and in guano. Guanine can be synthesized from any of the following methods:

i. Fischer's method (1897)

This method consists of synthesizing guanine from 2,6,8-trichloropurine, which on hydrolysis (aqueous KOH) gives 2,8-dichloro-6-oxo-purine. 2,8-Dichloro-6-oxo-purine was ammoniated with alcoholic ammonia, and converted to guanine on reduction with hydroiodic acid:

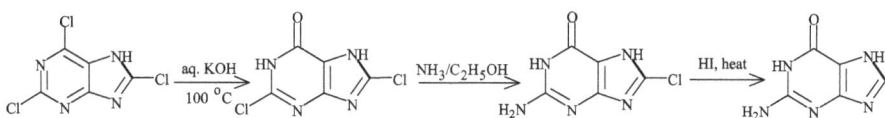

ii. Traube's synthesis (1900)

Traube's approach of guanine synthesis is entirely different from Fischer's method. He first synthesized the pyrimidine moiety of purine, converted it into diamino derivative, which was then cyclized to the imidazole. In this method, ethyl cyanoacetate was condensed with guanidine, which on cyclization yields 2,4-diamino-6-oxopyrimidine. This intermediate was treated with nitrous acid and reduced to give 2,4,5-triamino-6-oxopyrimidine, which on refluxing with formic acid was cyclized to guanine.

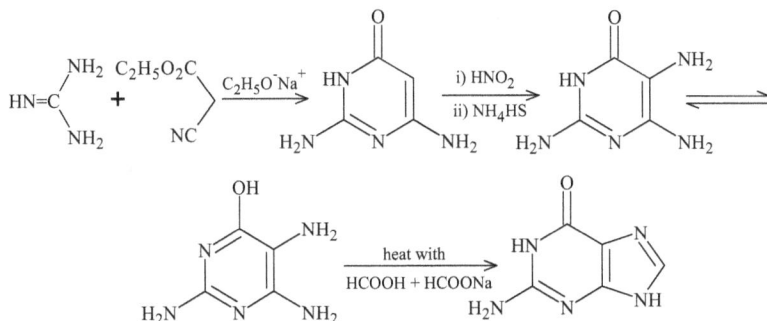

5.3.2 Pyrimidines

The term "pyrimidine" was coined by Pinner in 1885 and is derived from the words *pyridine* and *amidine*. It is one of the three diazines, having nitrogen atom at positions 1 and 3 of the ring, and the other two are pyridazine (nitrogen atom at positions 1 and 2) and pyrazine (nitrogen atom at positions 1 and 4). It occurs widely in nature as components of biologically important organic compounds such as vitamin B_1 (thiamine), vitamin B_2 (riboflavin), and alloxan. Pyrimidine (7) was first prepared by Gabriel and Colman in 1900, from barbituric acid (5) to 2,4,6-trichloropyrimidine (6) followed by reduction using zinc dust in hot water:

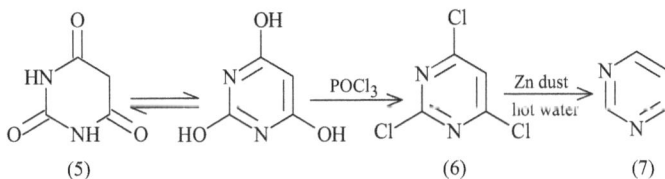

5.3.2.1 Thymine (2, 6-Dihydroxy-5-Methylpyrimidine, 5-Methyluracil)

Thymine was first isolated from calves' thymus glands by Albrecht Kossel and Albert Neumann in 1893. It is one of the four bases present only in DNA, whereas RNA uses uracil instead of thymine. Now, the question arises, why does DNA contain thymine and not uracil? The answer is thymine has greater resistance to photochemical mutation than uracil. Also, uracil can be easily formed internally on deamination of cytosine by specific enzymes, which alters the original number of uracil and hence repair systems would not be able to distinguish between original uracil from uracil generated from deamination of cytosine. As we have already discussed, DNA acts as a genetic material in almost all living organisms except few viruses; therefore, it is of utmost importance that its stability and integrity should be maintained. So, having thymine in DNA instead of uracil makes it more stable as genetic information should not be altered so easily. RNA, however, contains uracil because the instability does not matter for RNA as much because it is short-lived and gets degraded once it served its purpose.

Thymine can be prepared from the following methods:

i. Fischer and Roeder's method (1901)

This method involves the condensation between urea and ethyl methacrylate resulting in the formation of 5,6-dihydro-2,4-dioxo-5-methylpyrimidine (5,6-dihydro-5-methyluracil). The latter on treatment with bromine in acetic acid followed by boiling in pyridine yields thymine.

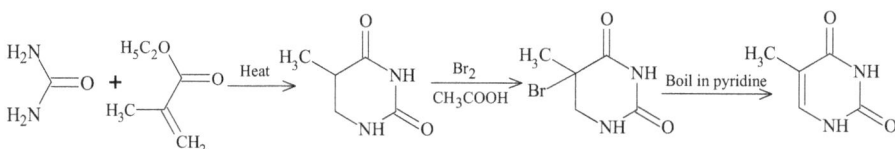

ii. Wheeler and Liddle (1908)

In this method, thiourea is condensed with sodium salt of formylpropionic ester to form 4-hydroxy-2-mercapto-5-methylpyrimidine (thiothymine). Thiothymine was then desulfurized with chloroacetic acid to yield thymine:

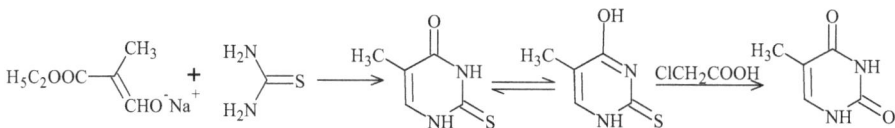

5.3.2.2 Cytosine (6-Amino-2-Hydroxypyrimidine)

Cytosine was first discovered from calf thymus tissues by Albrecht Kossel and Albert Neumann in 1894. A structure of cytosine was proposed in 1903, and was synthesized and confirmed in the same year. Cytidine triphosphate, an ester of cytosine and phosphoric acid, is utilized in the cells to react with nitrogen-containing alcohols to form coenzymes that play an important role in the formation of phospholipids.

The methods used for the synthesis of cytosine are as follows:

i. Wheeler and Johnson's method (1903)

The Wheeler and Johnson method involves the condensation of sodioformyl acetic ester with S-ethylisothiourea, resulting in the formation of 2-ethylmercapto-6-oxopyrimidine. Treatment of latter with $POCl_3$ yields the corresponding 4-chloro-2-ethylthiopyrimidine, which on subsequent reaction with alcoholic ammonia forms

4-amino-2-ethylthiopyrimidine. Treatment of 4-amino-2-ethylthiopyrimidine with dilute mineral acid resulted in the formation of 6-amino-2-oxopyrimidine (cytosine):

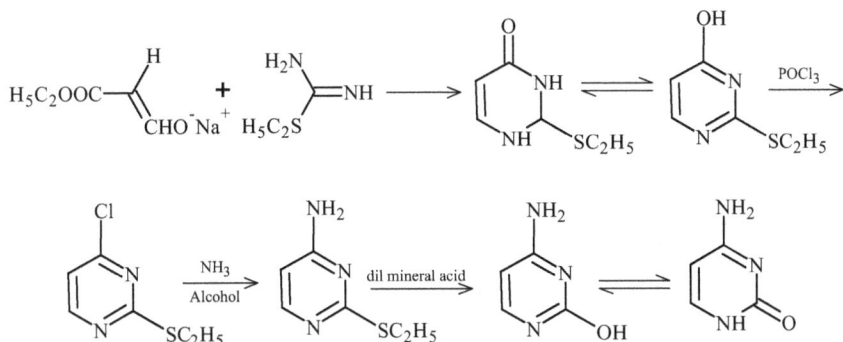

5.3.2.3 Uracil (2, 6-Dihydroxypyrimidine)

Uracil is a naturally occurring pyrimidine base having oxo group at 2- and 4-positions. The term "uracil" was coined by the German chemist Robert Behrend in 1885 while trying to prepare derivatives of uric acid. However, it was originally discovered and isolated in 1900 by Alberto Ascoli as a hydrolytic product of yeast nuclein. Uracil was also found in bovine thymus and spleen, and wheat germ. It has been synthesized in many ways:

i. Fischer and Roeder's method (1901)

This method involves condensation between urea and ethyl acrylate resulting in the formation of dihydrouracil, which reacts with bromine in acetic acid to form 5-bromo -5,6-dihydrouracil. The latter on boiling with pyridine yields uracil:

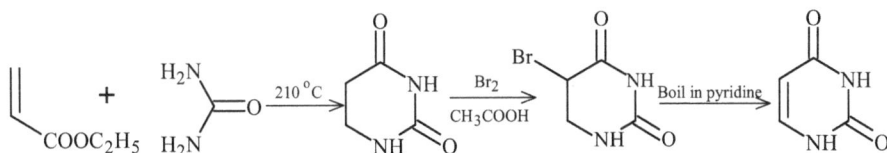

ii. Wheeler and Liddle's method (1908)

Uracil can also be obtained by reacting with thiourea and sodium salt of formyl acetic ester, and then boiling the product, 2-thiouracil, with chloroacetic acid:

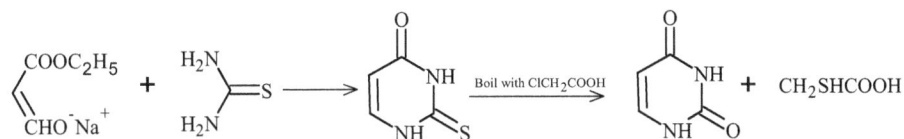

iii. Davidson et al.'s method (1926)

This approach involves dehydration of malic acid to formyl ethanoic acid which then reacts with urea to yield uracil:

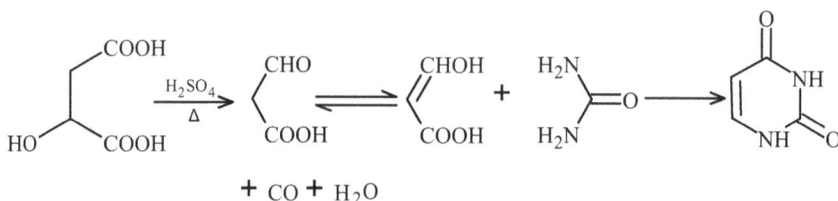

$$+ \ CO + H_2O$$

iv. Brown's method (1994)

Uracil can also be prepared by condensing maleic acid with urea in fuming sulfuric acid:

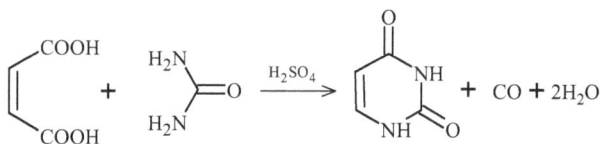

5.4 Structure of Polynucleotides

5.4.1 Primary Structure of DNA

The primary structure of DNA is the arrangement in which one nucleotide unit is joined to another nucleotide and describes the sequence of bases in a DNA strand. The 3′-hydroxyl (3′-OH) of the deoxyribose sugar of one nucleotide is joined to the phosphate group via ester linkage, which in turn joined to the 5′-OH of the adjacent sugar. The individual nucleotides in DNA and RNA are joined by the 3′–5′ phospho-diester bond, that is, the nucleotides are joined from the 3′-sugar carbon of one nucle-otide through the phosphate to the 5′-sugar carbon of the adjacent nucleotide. This is termed as 3′–5′ phosphodiester bond. The chain of the sugars joined by phospho-diester bonds is termed as the backbone of the nucleic acid. A schematic representa-tion of the primary structure of DNA is shown in Figure 5.6. RNA has a similar structure but it contains ribose sugar rather than deoxyribose and uracil in place of thymine. A single nucleic acid strand formed by phosphodiester bond has two chemi-cally distinct ends – namely 5′-end with phosphate group and 3′-hydroxyl group end. By convention, the sequence of bases is written in the 5′ to 3′ direction.

5' – GTCA – 3'

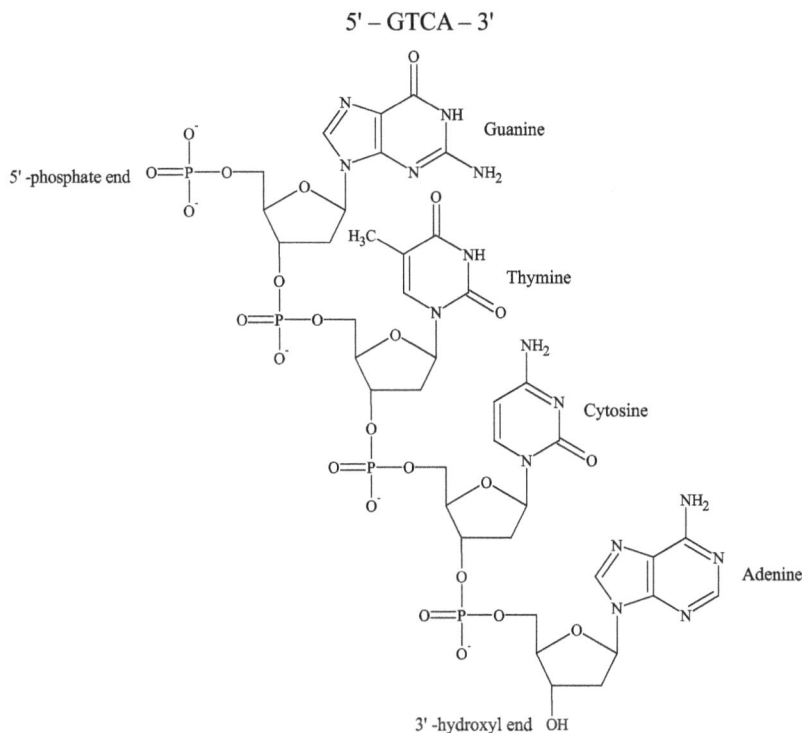

Figure 5.6: Primary structure of DNA.

5.4.2 Secondary Structure of DNA

DNA is a carrier of genetic information in humans and in almost all living organisms, except few viruses. Most of the DNA is present in the nucleus of the cell (nuclear DNA), but a small amount can also be found in the mitochondria (mitochondrial DNA). The size of the cell is very small, and organisms have many DNA molecules per cell, so each DNA molecule is tightly packed. In the nucleus, the DNA is wrapped around a set of proteins (called histones) and forms threads of chromatin, which further condenses to form chromosomes. Different organisms have different number of chromosomes, and human contains 23 pairs of chromosomes – 22 pairs of autosomes (numbered chromosomes) and 1 pair of sex chromosomes (X and Y). A single DNA–histone-associated complex is known as a nucleosome. Human DNA consists of 3 billion base pairs, and to study the physicochemical properties of DNA, it must be isolated and purified from the nucleus.

The double-helical model of DNA was discovered through the work of American molecular biologist James Watson, British physicist Francis Crick, British chemist Rosalind Franklin, and other researchers. Many people believe that Watson and Crick discovered the structure of DNA in the 1950s. However, other researchers have also contributed toward this noble and marvelous discovery. Following Miescher's discovery, other scientists – most notably, Phoebus Levene and Erwin Chargaff – performed a series of experiments, which revealed important information about the structure of DNA, for instance, the primary chemical components of DNA, and the way these components are attached to each other. Some key points which were noted by Chargaff's and his coworkers are:

(i) Similar base compositions exhibited by the closely related organism.
(ii) Identical base composition was observed for different cells and tissues of the same organism.
(iii) The base composition of DNA of an organism is characteristic of that organism.
(iv) The amount of adenine (A) equals the amount of thymine (T), and the amount of guanine (G) equals the amount of cytosine (C) in all species.

Rosalind Franklin made the most significant contribution to the discovery of DNA double helix. She discovered the density of DNA, and obtained images of DNA using X-ray crystallography, which established that the molecule exists in helical form. Without the scientific foundation laid by these pioneers, Watson and Crick would never have reached one of the biggest and groundbreaking discoveries – the three-dimensional double-helical model of DNA. James Watson, Francis Crick, and Maurice Wilkins got the Nobel Prize in 1962 for this model (Figure 5.7). Unfortunately, Franklin had an untimely death due to ovarian cancer in 1958 and Nobel committee does not confer the prize posthumously. The important features of the Watson–Crick model of DNA (known as double-helical model of DNA) deduced from the diffraction pattern are:

(i) Two polynucleotide chains run in opposite directions around a common axis. The chains are antiparallel meaning that if one strand is in 5′ → 3′ direction, then the other is in 3′ → 5′.
(ii) The sugar phosphate backbone is present on the outside of the double helix, and the purine and pyrimidine bases lie inside the helix.
(iii) The distance between the 2 bases (also known as axial rise) is 3.4 Å and the double-helical structure contains 10 base pairs per turn.
(iv) The diameter of the DNA double helix is 20 Å.
(v) The two strands of the double-stranded DNA are held together by the complementary base pairing – adenine is hydrogen bonded to thymine via two hydrogen bonds, whereas guanine pairs with cytosine via three hydrogen bonds (Figure 5.8).

Therefore, in a double helix, the number of adenine = number of thymine, and the number of guanine = number of cytosine. This is called Chargaff's rule. Erwin Chargaff in 1950 showed that ratios of guanine to cytosine and adenine to thymine are nearly the same in the DNA of any species and any organism.

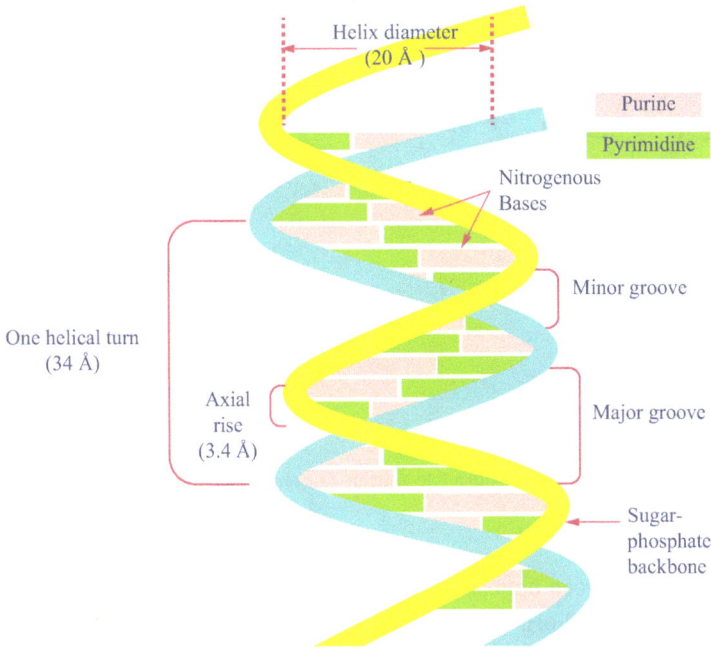

Figure 5.7: Watson–Crick double helix.

5.4.2.1 Factors Stabilizing DNA Double-Helical Structure

The structure of duplex DNA is stabilized by the balance of noncovalent forces in aqueous solution, and these forces are:

i. Hydrogen bonding

A hydrogen bonding arises from a dipole–dipole interaction between an electronegative atom and a hydrogen atom. It is one of the important factors that stabilize the secondary and tertiary structures of biomacromolecules like nucleic acids and proteins. In DNA, this bonding is specific; adenine always bonds to thymine via two hydrogen bonds and guanine always bonds to cytosine through three hydrogen bonds. The Watson–Crick hydrogen bonding pattern is shown in Figure 5.8.

ii. Stacking interaction

In DNA, π-stacking occurs between adjacent bases and adds to the stability of the molecular structure. The nitrogenous bases are aromatic and planar, and can stack over each other. Though a noncovalent bond is weaker than a covalent bond, the sum of all π-stacking interactions within the double-stranded (ds) DNA molecule creates a large net stabilizing energy.

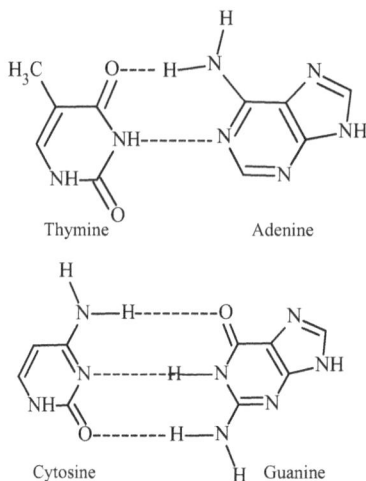

Figure 5.8: Watson–Crick hydrogen bonding.

iii. Ionic interaction

These interactions occur between oppositely charged species. Phosphodiester groups are fully ionized at physiological pH; thus, the exterior of the double helix carries two negative charges per base. The positively charged ions present in the cell such as Na^+, K^+, Mg^{2+}, or basic proteins interact with the negatively charged phosphate backbone, hence negating the electrostatic repulsion between the two backbones. Thus, the neutralization of negatively charged phosphate backbone with the help of salts and other ligands provides stability to DNA double helix.

5.5 Concept of DNA Duplex Formation and Its Characterization

The double-helical structure of DNA is remarkably stable and we have already discussed the various factors responsible for the stability of this miraculous biomolecule. DNA duplex stability is determined primarily by hydrogen bonding but base stacking and concentration of ions also play a significant role. A DNA duplex is formed by taking equimolar amounts of two complementary sequences of DNA (say Ass 5'-ATGGCCAATT -3' and Bss 3'-TACCGGTTAA-5', ss = single strand) in appropriate buffer and ionic conditions at physiological pH. On mixing at ambient temperature, the two strands of DNA will spontaneously form a duplex as illustrated in Figure 5.9. The degree of stabilization of a duplex depends on certain parameters like the number of guanine bases in a sequence, salt concentration, and nature of ions (monovalent or bivalent) present in the solution. The more the number of -GC- in a duplex, higher will be the stability of the duplex as guanine pairs with cytosine via three hydrogen bonds, whereas adenine forms only two hydrogen bonds with thymine. Base stacking is one among the important factor that stabilizes a double helix, and base-stacking interactions increase with

increasing salt concentration as high salt concentration decreases the repulsion between two negatively charged phosphodiester backbones. Also, divalent cations such as
magnesium (Mg^{2+}) are more stabilizing than monovalent cations (Na^+), and DNA duplex
stability therefore increases with increasing salt concentration but up to a certain concentration and in the presence of divalent cations.

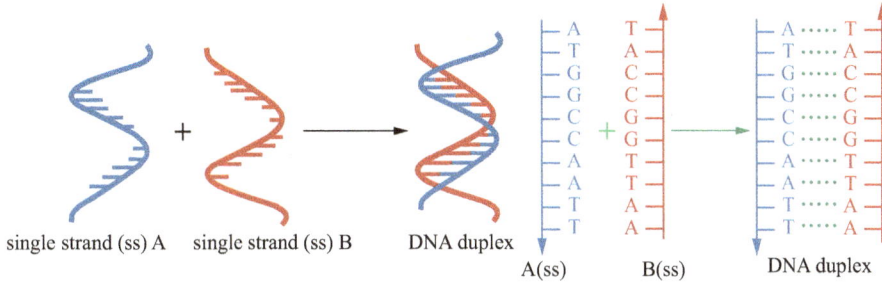

Figure 5.9: Formation of a double helix.

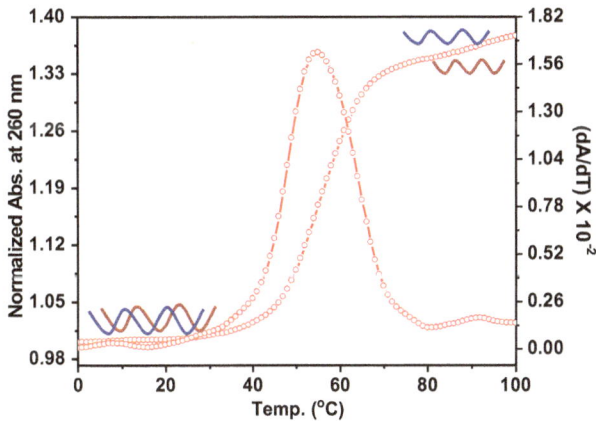

The stability of DNA duplex can be determined by UV (ultraviolet) melting/UV thermal
denaturation method. Increasing the temperature of DNA sample destabilizes the double helix, resulting in the separation of two strands. Heat disrupts the hydrogen bonding and hydration shell that leads to the loss of forces holding the two DNA strands
together. This process of separation of the two strands of DNA by increasing temperatures (heating) is known as thermal denaturation or melting, and on denaturation,
DNA separates into two single strands. The temperature at which 50% of ds DNA is
changed to ss DNA is called melting temperature or T_m. The melting temperature of a
DNA duplex depends on the base composition of DNA, and the greater the GC content
of the DNA, the higher the melting temperature.

DNA denaturation can be studied by different methods and one such method involves the measurement of absorption at 260 nm. Nucleic acids (DNA and RNA) absorb light in the UV region with a band centered at 260 nm. The absorption profile of DNA can be used to determine the concentration and purity of a DNA sample. ss DNA shows a 40% increase in absorbance in comparison to that of ds DNA. The increase in absorbance at 260 nm on heating DNA is called the hyperchromic shift (hyperchromicity) and is a result of unstacking of the bases. Rapid cooling of a heat-denatured DNA sample results in the formation of an unordered random coil, and under such conditions, two DNA strands cannot reform a perfect duplex. However, when the heated DNA sample is slowly cooled to room temperature, the separated strands can recombine and form the double helix. This process of slow cooling is known as renaturation or annealing.

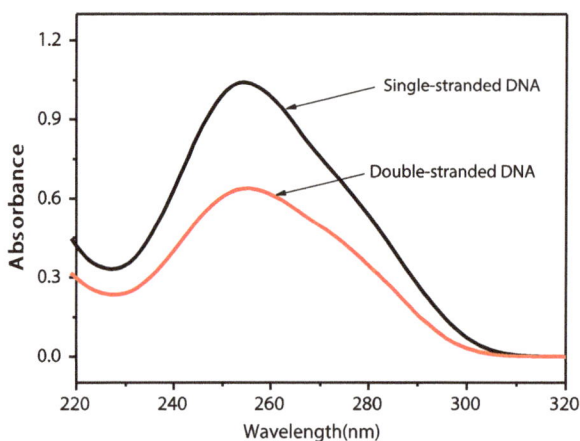

5.6 Biological Roles of DNA and RNA

DNA is a complex biological macromolecule that exhibits structural polymorphism, and is present in its supercoiled form in the cell of the organisms. This multifarious molecule plays important roles during essential biological processes. DNA maintains and carries the genetic information to each new cell through replication. This genetic information is then transcribed to messenger RNA (mRNA), and later transferred to the proteins through the process of translation. This concept was first given by Francis Crick in 1957 and is known as "central dogma of molecular biology" (Figure 5.10).

Like DNA, RNA is a long polymer made up of nucleotides joined by phosphodiester bond that carries out numerous functions in the cell and is found in all living organisms including viruses, plants, animals, and bacteria; however, there are two main differences. It has uracil in place of thymine, and sugar is ribose rather than deoxyribose as in DNA. Most RNA molecules adopt single strand conformation but if it contains base pairs that

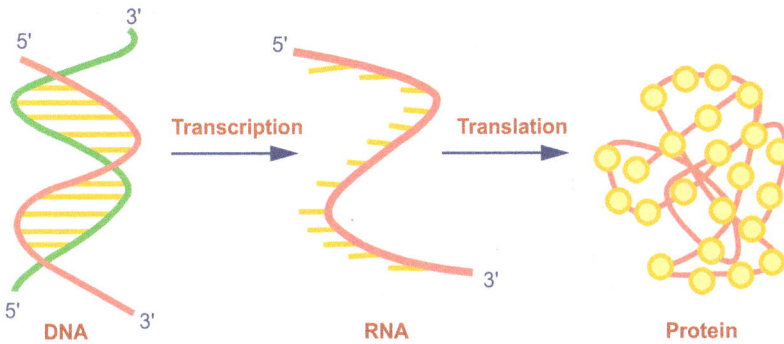

Figure 5.10: Central dogma of molecular biology.

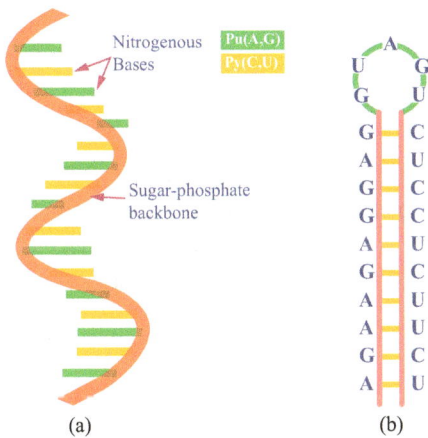

Figure 5.11: (a) Single-stranded RNA structure and (b) RNA forming an internal double-stranded structure (an example of self-complementarity in RNA).

can form complementary base pairing (A with U, and G with C), then RNA strand folds back on itself, and may adopt ds structure as shown in Figure 5.11(b).

There are various types of RNA: messenger RNA (mRNA), ribosomal RNA (rRNA), transfer RNA (tRNA), small nuclear RNA, microRNAs, small interfering RNA, and small nucleolar RNAs. Although all RNAs play a crucial role in the life processes of the cell; mRNA, rRNA, and tRNA are involved in protein synthesis.

i. rRNA: rRNA molecules are synthesized in the nucleolus and constitute 80% of the cellular RNA. Ribosomes, multicomponent complex, are made up of rRNA and ribosomal proteins, which act as the site for protein synthesis. Ribosomes are composed of two subunits: a large subunit and a small subunit. All eukaryotic 80S ribosomes comprise two subunits (60S and 40S), whereas all prokaryotic 70S ribosomes are made up of 50S and 30S (where S = Svedbergs, a unit used to measure how fast molecules move in a centrifuge). These complex structures move along the mRNA molecules during translation and facilitate the formation of polypeptide chains.

ii. mRNA: mRNA is formed from the DNA during transcription and carries genetic information from the nucleus to the cytoplasm where synthesis of proteins takes place. It is the least abundant of all RNAs and accounts for just 5% of the total RNA present in the cell. mRNA contains the base complementary to DNA and acts as the template for protein synthesis, the only difference being thymine is replaced by uracil. The sequence of bases in mRNA determines the order of amino acids to be inserted in a polypeptide chain. In rapidly growing cells, different types of proteins are required to perform various crucial functions in the body within a short time interval; therefore, fast turnover in protein synthesis becomes essential. So, mRNA is formed as per the requirement and gets degraded once it performed its function so that the nucleotides can be recycled.

iii. tRNA: tRNA molecules play an important role in protein synthesis and are located in the cellular cytoplasm. These are small molecules which contain 74–95 nucleotides and have a cloverleaf secondary structure formed by internal complementary base pairing (Figure 5.12). tRNAs work as a physical link between the mRNA and the sequence of amino acid in a polypeptide chain. It picks up a specific amino acid from the cytoplasm and attaches it to the ribosome. Complementation between a 3-nucleotide codon in mRNA and 3-nucleotide anticodon of the tRNA results in the synthesis of proteins based on the sequence specified by mRNA.

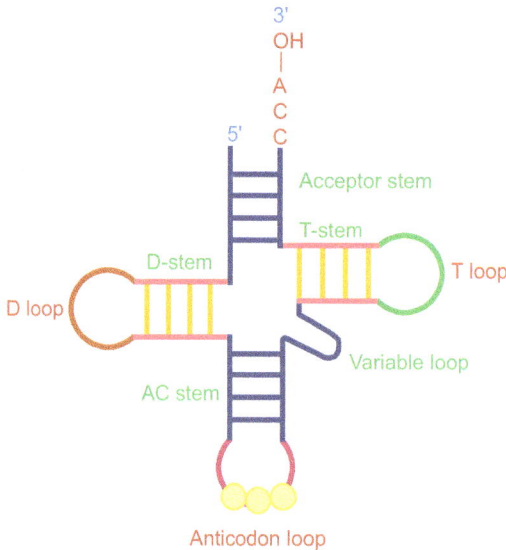

Figure 5.12: Cloverleaf secondary structure of tRNA.

RNA is less stable than DNA because of the presence of hydroxyl "-OH" group at 2'-position. Unlike DNA, RNA can be easily cleaved because the 2'-OH of ribose can act as a nucleophile and attack the adjacent phosphodiester group (Figure 5.13). This reaction does not take place in DNA because it lacks 2'-OH group and thus remain intact throughout the life span of a cell. RNA, in contrast, is synthesized when it is needed and is degraded, once it has served its purpose.

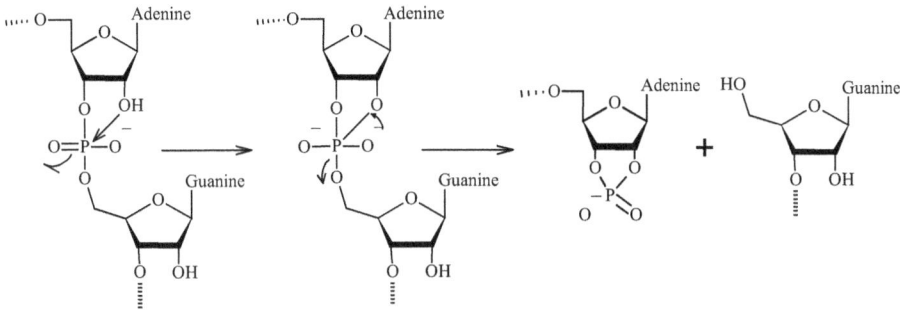

Figure 5.13: RNA is less stable than DNA.

5.7 DNA Replication

DNA replication is an important biological process that occurs in all living organisms during cell division. In this process, DNA makes a copy of itself, which is the basis for biological inheritance. The first step in this process is the unwinding of the DNA double helix to create single strand of DNA. Each strand can act as a template for the synthesis of new strand (daughter strand). This process is also known as semiconservative mode of replication because each ds DNA formed as a result of replication contained one of the original strands and one new strand, displayed in Figure 5.14.

In a cell, replication is initiated at specific locations in the genome, known as "origins." Origins are "AT-rich sequences" (rich in adenine and thymine bases) which assist this process, as these strands (AT-rich) can be easily separated as more energy is needed to break G–C hydrogen bonding.

Unwinding of DNA starts at the origin; the two strands are separated and form a replication fork. The process of DNA replication is catalyzed by different enzymes which perform crucial roles as given in Table 5.2.

In DNA replication, the DNA strands are separated by DNA helicase into two template strands. RNA primers are generated on the parent strands, and RNA primase then adds primers to the template strands. DNA polymerase III then synthesizes the new DNA strands by adding nucleotides in a 5'→3' direction (Figure 5.15). Deoxyribo-NTPs (dNTPs) are precursors for the DNA synthesis, and two of the phosphate groups

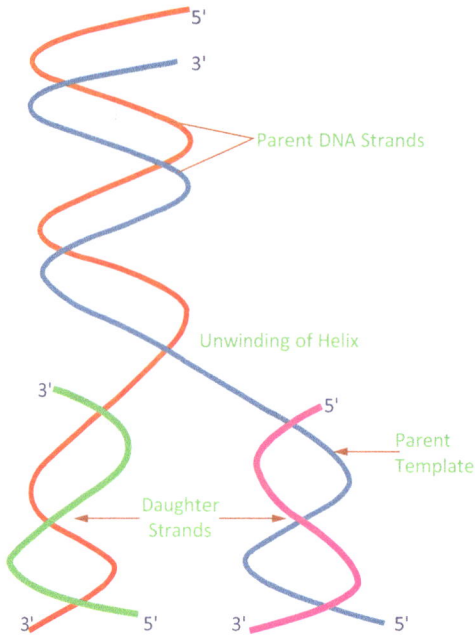

Figure 5.14: Semiconservative mode of replication.

Table 5.2: List of enzymes involved in DNA replication.

Enzyme	Function in DNA replication
Topoisomerase	Relaxes the DNA from its super-coiled nature.
DNA helicase	Unwinds the DNA double helix by breaking the hydrogen bonds. This enzyme is also known as helix-destabilizing enzyme.
DNA polymerase	Synthesizes a new DNA strand by adding nucleotides in the 5' to 3' direction. This class of enzyme also carries out the proofreading and error correction.
Single-strand binding (SSB) proteins	Bind to single-strand DNA (ssDNA), stabilize unwound DNA, and prevent it from forming double helix.
DNA ligase	Joins the Okazaki fragments of the lagging strand.
RNA primase	Enzyme that creates a RNA primer on the parent strand to initiate the synthesis of a new DNA strand.

break off during the replication process to release energy. Since the two strands are antiparallel to each other and the synthesis can only occur in 5'→3' direction, one of the strands can be synthesized in a continuous manner but the other strand has the opposite orientation, that is, 3' → 5'; so, the synthesis of this strand takes place in a

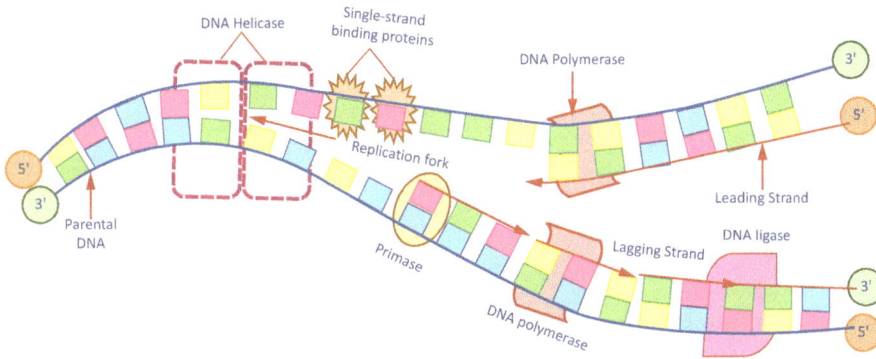

Figure 5.15: Replication of DNA.

discontinuous manner, that is, in small fragments. These small fragments are called Okazaki fragments. The one strand which is synthesized continuously is called leading strand, whereas the other one is termed as lagging strand whose synthesis occurs in small fragments. The leading strand needs only one primer whereas lagging strand requires more primers as it is synthesized in small fragments. Another DNA polymerase (DNA polymerase I) removes RNA primers and replaces them with DNA. Okazaki fragments are then joined together by DNA ligase, thus completing the synthesis of newly replicated DNA strands.

DNA replication begins at many points in eukaryotic chromosomes; hence, initiation and termination of replication forks occur at many points in the chromosome, and these are not controlled in a particular way. The progress of the DNA replication fork must stop/block to terminate replication. There are two methods which organisms can adopt. First is to have a termination site sequence in the DNA, and the second, it can be physically stopped by a protein that can recognize and bind to the sequence. This function is achieved by *Ter* protein, a DNA replication terminus site-binding protein. In case of bacteria, the replication process terminates when the two replication forks meet on the opposite end of the same stretch of DNA and fuse. This occurs in the terminus region, which is situated completely opposite the origin, for example, in *Bacillus subtilis* and *Escherichia coli*.

5.8 Biosynthesis of RNA: Transcription

The transformation of information of DNA into functional molecules is termed as gene expression. The biosynthesis of RNA from DNA template is called transcription, and this biological process is catalyzed by the enzyme RNA polymerase. Transcription takes place in the nucleus of the cell; newly synthesized RNA then leaves the nucleus and moves into the cytoplasm. RNA polymerases are very large and complex enzymes which perform various functions in this process. The synthesis of RNA synthesis takes place in three stages:

(i) initiation

(ii) elongation

(iii) termination

5.8.1 Initiation

The process of transcription begins when RNA polymerase searches DNA for some specific sites, known as promoter sites from within large stretches of DNA. The promoter sites are present at the beginning of genes. Eukaryotic genomes consist of promoter sites having TATAAA consensus sequence, called *TATA box* or *Hogness box*, centered at −25, shown in Figure 5.16(a). Many eukaryotic promoters also have a *CAAT box* with a GGNCAATCT (where N is any nucleotide) consensus centered at −75. In bacteria, two sequences which work as promoter sites are present on the 5′-(upstream) side of the first nucleotide to be transcribed. One is *Pribnow box*, centered at −10 with a consensus sequence of TATAAT, and the other having a consensus sequence TTGACA is centered at −35 region (Figure 5.16(b)).

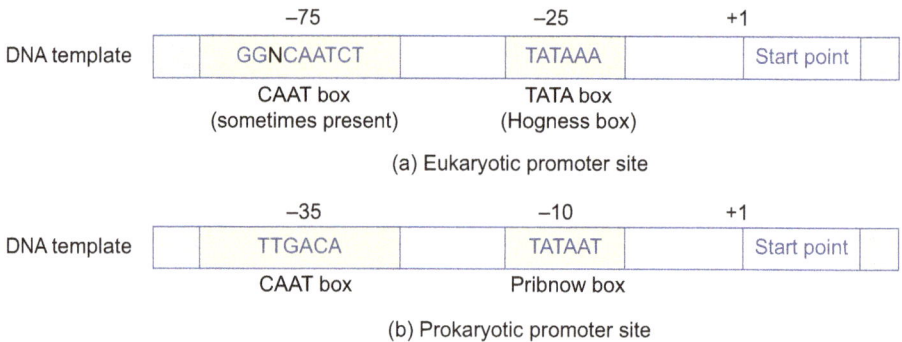

	−75		−25		+1	
DNA template	GGNCAATCT		TATAAA		Start point	

CAAT box (sometimes present)   TATA box (Hogness box)

(a) Eukaryotic promoter site

	−35		−10		+1	
DNA template	TTGACA		TATAAT		Start point	

CAAT box   Pribnow box

(b) Prokaryotic promoter site

Figure 5.16: Promoter sites for transcription in (a) eukaryotes and (b) prokaryotes.

RNA polymerase catalyzes the initiation as well as the elongation of RNA chains. They unwind a short stretch of ds DNA to form ss DNA. The DNA template strand is read in the 3′→5′ direction so that RNA can be synthesized in the 5′→3′ direction. The bases in the DNA template strand determine the sequence of the bases that need to be incorporated into mRNA strand. In contrast to DNA synthesis, RNA synthesis does not require any primer for the initiation process.

5.8.2 Elongation

The elongation phase of transcription starts with the formation of the first phospho-diester bond. RNA polymerase picks the correct ribo-NTPs (rNTPs) and catalyzes the formation of a phosphodiester bond. Since the RNA chain grows in the 5′→3′ direction, the 3′-hydroxyl group of the RNA is free to attack the α-phosphorus atom of next rNTP. The polymerization reaction catalyzed by a RNA polymerase which takes place within a complex DNA is termed as "transcription bubble" (Figure 5.17). This process continues as the enzyme moves along the DNA template, and is catalyzed by a single RNA polymerase from start to end. The region within the transcription bubble con-sists of RNA polymerase, DNA, DNA–RNA hybrid helix, and nascent RNA. The forma-tion of hybrid double helix takes place between newly synthesized RNA and DNA.

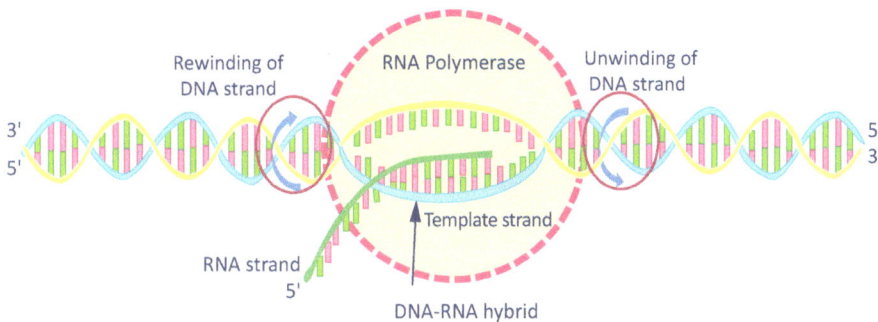

Figure 5.17: A schematic representation of transcription bubble.

5.8.3 Termination

The termination phase of transcription involves the breaking of phosphodiester bonds and dissociation of RNA–DNA hybrid. The rewinding of DNA strands takes place with the release of newly synthesized RNA strand. How the process of transcrip-tion terminates? Termination of transcription takes place in two ways:

i. *Rho (ρ)-independent transcription termination:* This type of termination takes place when palindromic GC-rich sequences are present in RNA transcript followed by an AT-rich region. GC-rich sequences favor the formation of a hairpin structure (contains loop and stem part), which is followed by four or more uracil residues. The uracil bases are quite important for termination. RNA polymerase stops immediately after synthesizing a stretch of RNA which forms a hairpin structure, and RNA–DNA helix hybrid formed after the hairpin structure is unstable because of weak base pairing

between rU and dA. So, this pause in transcription bubble instigated by hairpin formation allows the nascent RNA to detach from the DNA template and then from RNA polymerase. Rejoining of two separated DNA strands occurs to form a double helix, and
transcription bubble closes. Hence, the rho-independent transcription termination occurs as shown in Figure 5.18.

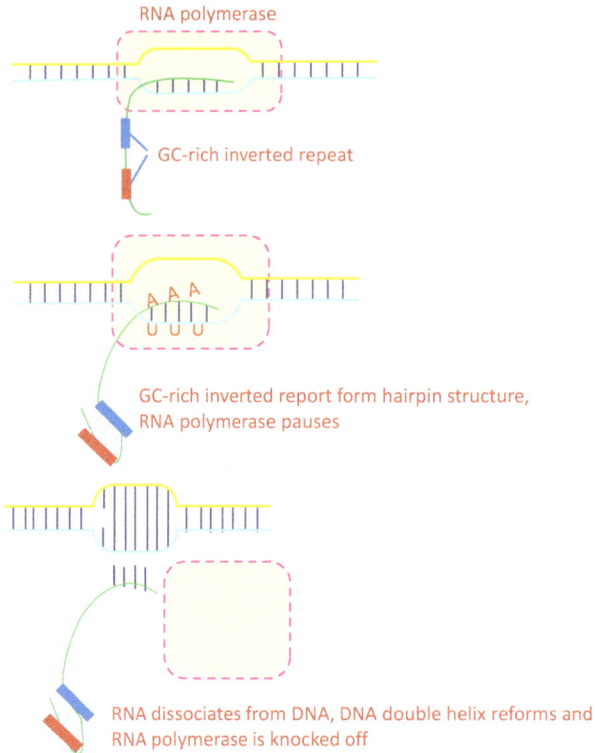

Figure 5.18: Rho (ρ)-independent transcription termination.

ii. *Rho (ρ)-dependent transcription termination:* RNA polymerase requires no additional
factors to terminate transcription, when GC-rich sequences followed by several uracils
are present in the newly synthesized RNA transcript, whereas additional factors are required at other sites to terminate transcription. In this method of termination, "Rho (ρ)"
protein destabilizes the DNA–RNA hybrid helical structure. Rho protein is an ATP-
dependent helicase that binds to the ss RNA that is rich in cytosine and poor in guanines
but not to the duplex DNA or duplex RNA. When ρ reaches RNA polymerase at the transcription bubble, it breaks off the DNA–RNA bonds, and pulls nascent RNA away
from RNA polymerase (Figure 5.19), hence terminating the transcription.

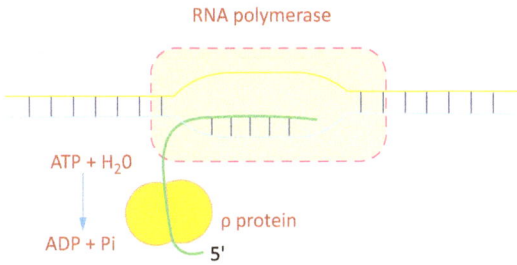

Figure 5.19: Termination of transcription by Rho (ρ) protein.

5.9 Biosynthesis of Proteins: Translation

The process of synthesis of proteins from mRNA is called translation. During this biological process, mRNA sequence is read using the genetic code. Genetic code is a set of rules that describe how mRNA sequence is converted into amino acids, which are the building blocks of proteins. As described earlier, DNA acts as a carrier of genetic information in almost all living organisms, except few viruses, where RNA constitutes the genetic material. Studies revealed that the synthesis and structure of proteins (polypeptides) are controlled by the genes. Now the question arises: how the sequence of four bases controls the sequence of 20 amino acids? During translation, 20 amino acids are incorporated in a polypeptide chain, which means the formation of 20 different codons is possible making use of 4 bases present in RNA. Obviously, there cannot be a 1:1 correspondence between mRNA bases and amino acids, and 2:1 correspondence was inadequate. If there are three bases per amino acid, then the number of possible triplets $= 4^3$ (64), which would be enough to code for each amino acid. So, the sequence of three nucleotides was termed as codon, which specifies a single amino acid. The first evidence for the presence of three nucleotides per codon came from the experiment performed by Crick and colleagues in 1961. Francis Crick, Sydney Brenner, Leslie Barnett, and Richard Watts-Tobin first demonstrated the three bases of DNA code for one amino acid. However, Marshall Nirenberg and Heinrich Matthaei "cracked" the first "word" of the genetic code. Starting with the pioneer work of Nirenberg and Matthaei, followed by Nirenberg and Philip Leder, the decoding was completed by Har Gobind Khorana in 1966. The deciphering of code, that is, assigning amino acid to each of the codons, was validated using various techniques (Table 5.3). The Nobel Prize in Physiology or Medicine 1968 was awarded jointly to Marshall W. Nirenberg, Har Gobind Khorana (who mastered the synthesis of nucleic acids), and Robert W. Holley (who discovered the chemical structure of transfer RNA) "for their interpretation of the genetic code and its function in protein synthesis."

Table 5.3: The genetic code.

First-letter 5′ end	Second letter								Third-letter 3′ end
	U		C		A		G		
U	UUU UUC	Phe	UCU UCC UCA UCG	Ser	UAU UAC	Tyr	UGU UGC	Cys	U C
	UUA UUG	Leu			UAA UAG	**STOP** **STOP**	UGA UGG	**STOP** His	A G
C	CUU CUC CUA CUG	Leu	CCU CCC CCA CCG	Pro	CAU CAC	His	CGU CGC	Arg	U C
					CAA CAG	Gln	CGA CGG		A G
A	AUU AUC AUA	Ileu	ACU ACC ACA ACG	Thr	AAU AAC	Asn	AGU AGC	Ser	U C
	AUG	Met **START**			AAA AAG	Lys	AGA AGG	Arg	A G
G	GUU GUC GUA GUG	Val	GCU GCC GCA GCG	Ala	GAU GAC	Lys	GGU GGC	Gly	U C
					GAA GAG	Glu	GGA GGG		A G

Amino acid names			
Ala = Alanine	Gln = Glutamine	Leu = Leucine	Ser = Serine
Arg = Arginine	Glu = Glutamate	Lys = Lysine	Thr = Threonine
Asn = Asparagine	Gly = Glycine	Met = Methionine	Trp = Tryptophan
Asp = Aspartate	His = Histidine	Phe = Phenylalanine	Tyr = Tyrosine
Cys = Cysteine	Ile = Isoleucine	Pro = Proline	Val = Valine

The synthesis of proteins takes place on ribosomes in the cytoplasm and occurs in four stages: activation, initiation, elongation, and termination. During translation, mRNA sequence is read using the genetic code. The genetic code is a set of three-letter combination of nucleotides called codons, and each of which corresponds to a specific amino acid or stop signal. Protein synthesis is catalyzed by rRNA molecules rather than enzymes. RNA molecules, found in ribosome, that act as catalysts are known as ribozymes. Ribosomes comprise a small and large subunit that environs the mRNA, and have three tRNA binding sites, named as the A (aminoacyl) site, the P (peptidyl) site, and E (exit) site. Prokaryotes and eukaryotes have different set of small and large subunits. A prokaryotic ribosome (70S) consists of a large 50S subunit and a smaller 30S subunit. A eukaryotic ribosome has 60S and 40S subunits; together they form 80S.

During the activation process, a link between the correct amino acid to its corresponding tRNA is catalyzed by amino-acyl tRNA synthetase. The amino acid is attached by its carboxyl group to the 3'-OH of the tRNA via ester bond, as demonstrated in Figure 5.20. The initiation process begins when the small subunit of the ribosome binds to 5'-end of mRNA with the help of initiation factors. The chain elongates when the next aminoacyl–tRNA (charged tRNA) binds to the ribosome with an elongation factor (EF) and guanosine triphosphate (GTP). Termination of the process occurs when a stop codon (UAA, UAG, or UGA) is found at the A site of the ribosome as these codons cannot be recognized by tRNA; rather releasing factor recognizes these codons and releases the polypeptide chain. The various steps involved in the protein synthesis are discussed further:

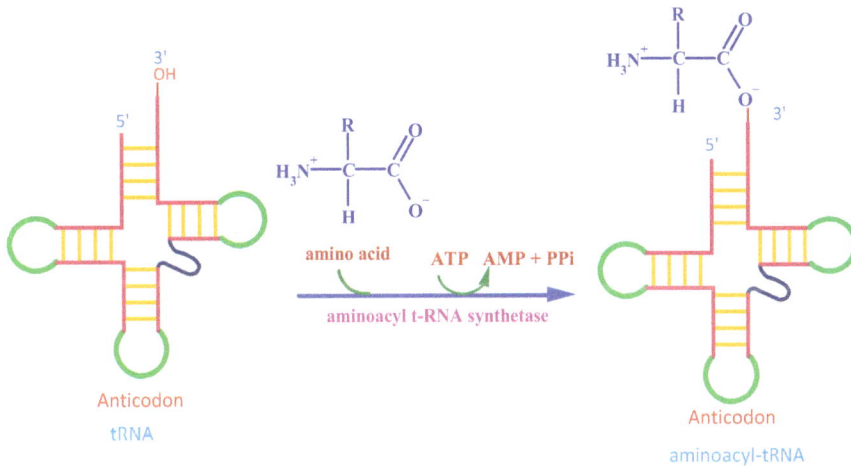

Figure 5.20: Attachment of an amino acid to a tRNA.

5.9.1 Initiation

– Initiation begins with the binding of a small subunit of the ribosome to the "upstream" region (on the 5'-side) of the start of the mRNA.
– It moves downstream (5' → 3') until it meets the start codon, AUG, which codes for methionine. Then, a large subunit and a special initiator tRNA join the small subunit on the ribosome.
– In eukaryotes, the initiator tRNA that carries methionine (Met) binds to the P (peptidyl) site on the ribosome, whereas bacteria use a modified methionine, fMet.

The steps involved in initiation of translation are shown in Figure 5.21.

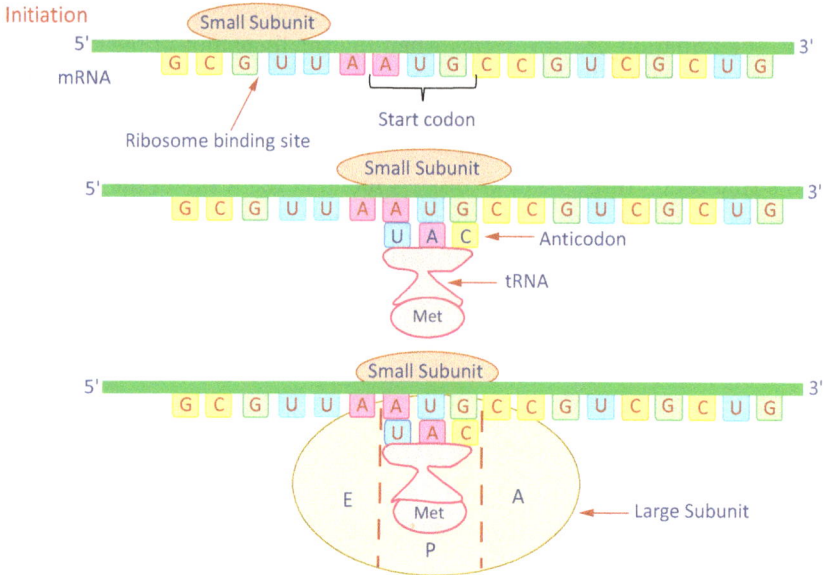

Figure 5.21: Steps involved in the initiation of translation.

5.9.2 Elongation

The elongation step of the polypeptide chain consists of three steps: (a) Aminoacyl–tRNA binding, (b) formation of peptide bond, and (c) translocation.

(a) Aminoacyl–tRNA binding: In this step of elongation cycle, the corresponding aminoacyl–tRNA (tRNA carrying the amino acid) base pairs with the next codon via codon–anticodon interaction at the A site on the mRNA. Binding of aminoacyl–tRNA requires EF (named EF-1 in eukaryotes, and EF-Tu in bacteria) and the source of the energy, GTP.

(b) Formation of peptide bond: In the second step, peptide bond formation takes place, which is catalyzed by peptidyl transferase. The amino acid bound to the tRNA via carboxyl end in the P-site decouples from the tRNA, and forms peptide bond to the amino group of the incoming amino acid attached to tRNA in the A-site.

(c) Translocation: Three concerted steps occur collectively in this step of elongation cycle. The ribosome shifts one codon downstream, and with this shifting, the more recently-arrived tRNA with associated peptide moves to the P-site, and leaves the A-site for the incoming new aminoacyl–tRNA. A complex of EF (called EF-G (translocase) in bacteria; EF-2 in eukaryotes) and GTP is required for this step.

P-site is the site where the growing polypeptide chain is bound. The A-site is so named because the incoming aminoacyl–tRNA binds to the A-site, that is, the tRNA bringing the next amino acid. The Exit site (or E-site) is a binding site, where tRNA goes prior to its release from the ribosomes.

5.9.3 Termination

- Termination step occurs when one of STOP codons (UAA, UAG, UGA, also called termination codons) becomes positioned in the A-site.
- Unlike other codons, there are no aminoacyl–tRNA molecules complementary to these STOP codons. However, they are recognized by protein release factors (RF) when they arrive at the A-site.
- In prokaryotes, RF-1 recognizes UAA and UAG, whereas RF-2 recognizes UGA. The third RF, RF-3, is also needed to assist RF-1 and RF-2. Eukaryotes have just one RF (eRF-1) that recognizes all the termination codons. eRF-3 may assist eRF-1 to terminate translation.
- In the next step, the RFs make the peptidyl transferase to transfer the synthesized polypeptide chain to a water molecule. This cleaves the bond between tRNA and the polypeptide in the P-site. The polypeptide now leaves the ribosome, followed by the mRNA and free tRNA, and the ribosome dissociates into 30S and 50S subunits ready to start translation afresh.
- The release of the polypeptide chain followed by the mRNA and free tRNA from the ribosome takes place in the final step, and the ribosome dissociates into its subunits to start the next round of translation.

All the steps involved in the elongation and termination of transcription are shown in Figures 5.22 and 5.23, respectively.

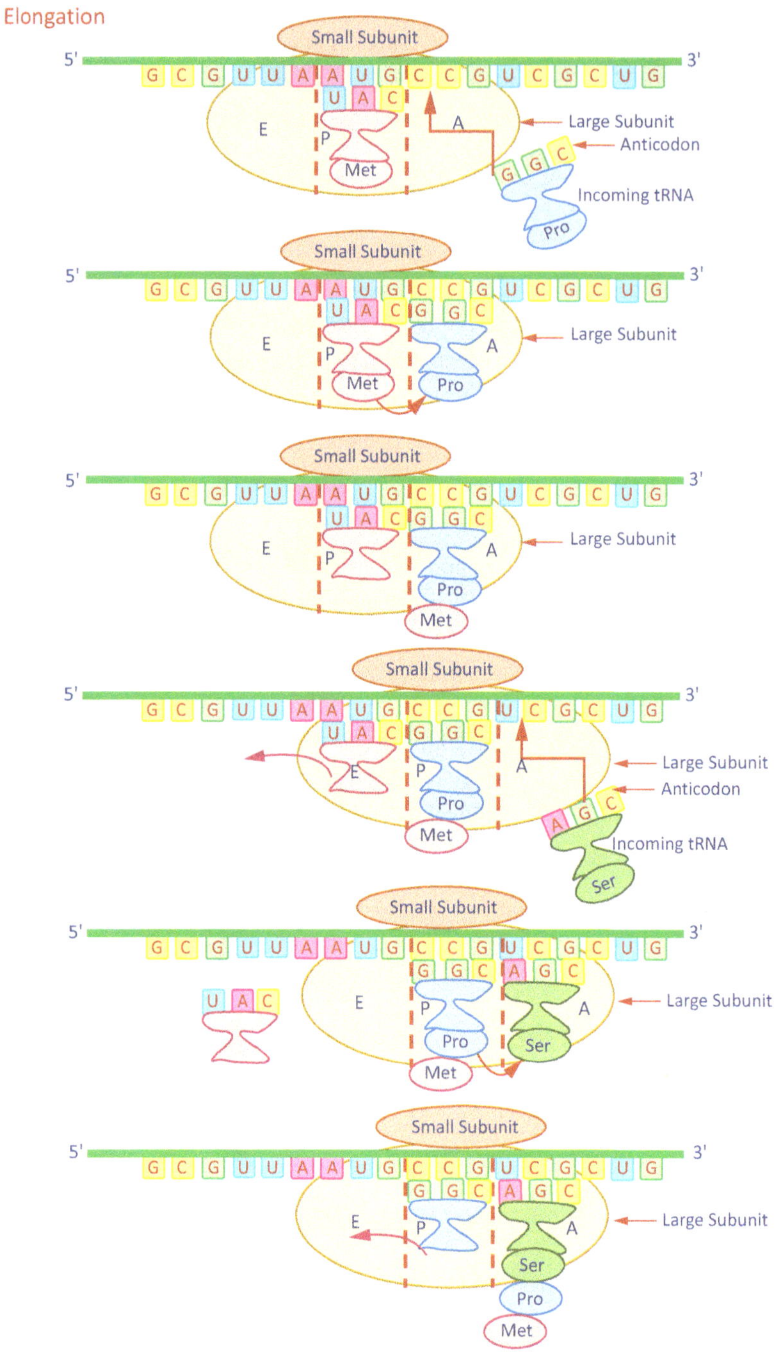

Elongation

Figure 5.22: Steps involved in the elongation of translation.

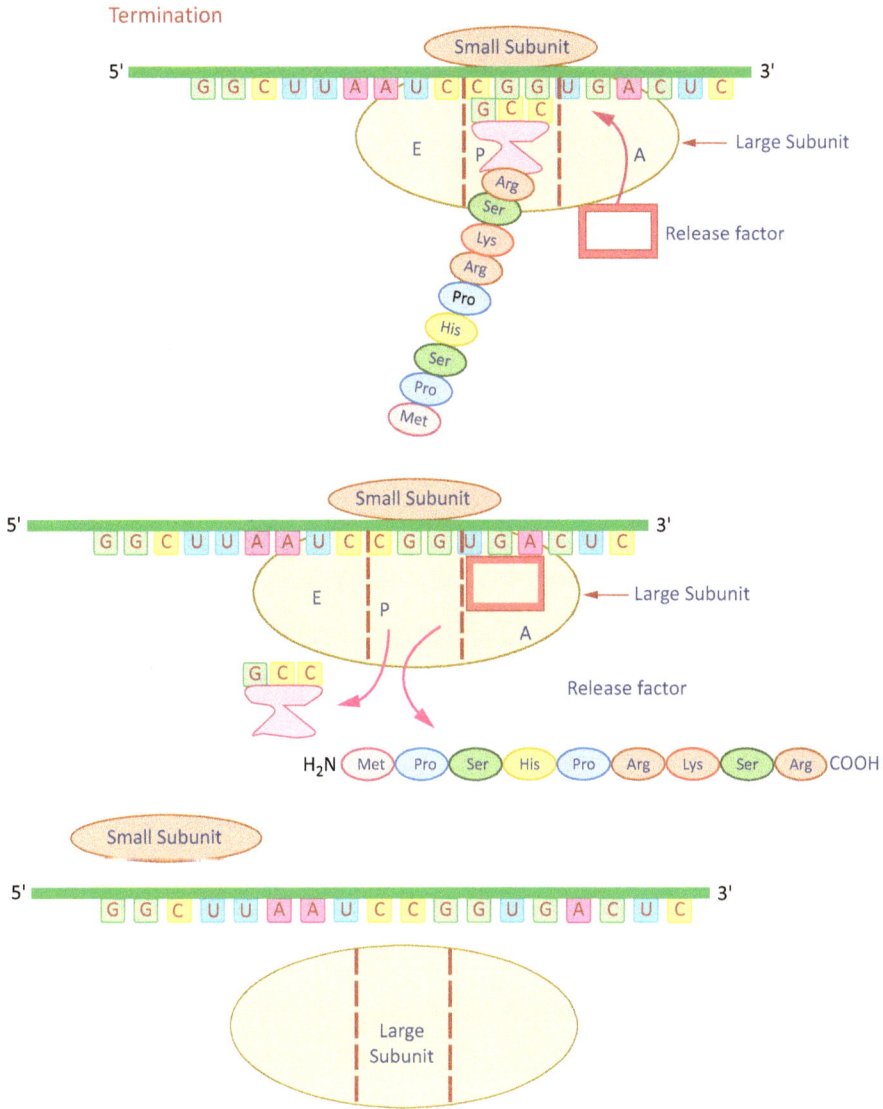

Figure 5.23: Steps involved in the termination of translation.

References

[1]   Berg JM, Tymoczko JL, Gatto GJ. Jr Lubert Stryer. Biochemistry. Freeman Macmillan. 8th Edition. ISBN-13. 2006:978–1.
[2]   Bruice, P. Y. Organic Chemistry, The Chemistry of the Nucleic Acids. pp 1207–1234. 8th ed.; Zalesky, J., Ed.; Pearson Education: London, 2016.
[3]   Bryce CF, Pacini D. The structure and function of nucleic acids. Biochemical Society; 1994.
[4]   Finar IL. Organic Chemistry (Volume 2), Dorling Kindersley (India) Pvt. Ltd. 1964.
[5]   Fischer E. Syntheses in the purine and sugar group. Nobel Lecture. 1902 Dec. Fischer E. Nobel Lectures, Chemistry 1901–1921. Amsterdam: Elsevier; 1966. Syntheses in the purine and sugar group; pp. 21–35.
[6]   Hames D, Hooper N. Instant Notes Biochemistry. Taylor & Francis Oxfordshire, United Kingdom; 2006 Sep 27.
[7]   http://www.acs.org/content/acs/en/education/whatischemistry/landmarks/geneticcode.html
[8]   Lehninger AL, Nelson DL, Cox MM. Lehninger Principles of Biochemistry. Fourth ed. New York: W.H. Freeman; 2005.
[9]   Saenger W. Defining Terms for the Nucleic Acids. In Principles of Nucleic Acid Structure Springer, New York, NY. 1984; 9–28.
[10] Sinden, R. R. DNA Structure and Function Academic Press, San Diego, 1994.
[11] Ullmann F. Ullmann's Biotechnology and Biochemical Engineering. Wiley-VCH; 2007. Burtscher, H.; Berner, S.; Seibl, R.; Muhlegger, K. Nucleic acids. In Ullmann's Biotechnology and Biochemical Engineering; Wiley-VCH: Weinheim, Germany, 2007; pp. 157–194.

Index

https://doi.org/10.1515/9783110793765-006

www.ingramcontent.com/pod-product-compliance
Lightning Source LLC
Chambersburg PA
CBHW081522220326
41598CB00036B/6294